# THE POLITICS OF EVOLUTION

The controversy over teaching evolution or creationism in American public schools offers a policy paradox. Two sets of values—science and democracy—are in conflict when it comes to the question of what to teach in public school biology classes. Prindle illuminates this tension between American public opinion, which clearly prefers that creationism be taught in public school biology classes versus the democratic ideal held by virtually every person in the country. An elite consisting of scientists, professional educators, judges, and business leaders by and large are determined to ignore public preferences and teach only science in science classes despite the majority opinion to the contrary. So how have the political process and the Constitutional law establishment managed to thwart the people's will in this self-proclaimed democracy?

Drawing on a vast body of work across the natural sciences, social sciences, and humanities, Prindle explores the rhetoric of the evolution issue, explores its history, examines the nature of the public opinion that causes it, evaluates the Constitutional jurisprudence that upholds it, and explains the political dynamic that keeps it going. This incisive analysis is a must-read in a wide range of disciplines and for anyone who wants to understand the politics of biology.

**David F. Prindle** is a professor in the Department of Government at the University of Texas at Austin. He has published research in the areas of voting and parties, energy policy, the presidency, and the politics of the entertainment media.

"In this eminently readable and thoroughly researched volume, political scientist David Prindle addresses the political implications of evolution, and particularly the debate over the teaching of evolution and creationism, including intelligent design, in public schools. Prindle argues convincingly that politics and science are inextricably intertwined; his thought-provoking message will spark discussion in the evolutionary sciences, history, philosophy, religion, and education, and should be read by anyone interested in the evolution–creationism controversy."
*Dr. Patricia H. Kelley, University of North Carolina Wilmington*

# THE POLITICS OF EVOLUTION

David F. Prindle

NEW YORK AND LONDON

First published 2015
by Routledge
711 Third Avenue, New York, NY 10017

and by Routledge
2 Park Square, Milton Park, Abingdon, Oxon, OX14 4RN

*Routledge is an imprint of the Taylor & Francis Group, an informa business*

© 2015 Taylor & Francis

The right of David F. Prindle to be identified as author of this work has been asserted by him in accordance with sections 77 and 78 of the Copyright, Designs and Patents Act 1988.

All rights reserved. No part of this book may be reprinted or reproduced or utilised in any form or by any electronic, mechanical, or other means, now known or hereafter invented, including photocopying and recording, or in any information storage or retrieval system, without permission in writing from the publishers.

*Trademark notice*: Product or corporate names may be trademarks or registered trademarks, and are used only for identification and explanation without intent to infringe.

*Library of Congress Cataloging in Publication Data*

Prindle, David F. (David Forrest), 1948–
 The politics of evolution / David Prindle.
  pages cm
 Includes bibliographical references and index.
 1. Evolution (Biology)—Political aspects. 2. Evolution (Biology)—Religious aspects 3. Evolution (Biology)—Public opinion. 4. Darwin, Charles, 1809–1882—Influence. I. Title.
 QH366.2.P743 2015
 576.8—dc23
 2014042230

ISBN: 978-1-138-88783-1 (hbk)
ISBN: 978-1-138-88784-8 (pbk)
ISBN: 978-1-315-71389-2 (ebk)

Typeset in Bembo
by Apex CoVantage, LLC

Printed and bound in the United States of America by Publishers Graphics, LLC on sustainably sourced paper.

*For Matthew, Caulette, Lylli, and Ivy. The future.*

# CONTENTS

| | |
|---|---|
| *List of Tables* | *ix* |
| *Acknowledgments* | *xi* |
| Introduction: *The Politics of Evolution* | 1 |
| 1   Biology: The Most Political Science | 8 |
| 2   Evolution and Metaphor | 33 |
| 3   Evolution and Religion | 54 |
| 4   Evolution and Public Opinion | 89 |
| 5   The Jurisprudence of Evolution | 116 |
| 6   Evolution and the Party Battle | 137 |
| *Bibliography* | *165* |
| *Index* | *181* |

# TABLES

| | | |
|---|---|---|
| 3.1 | Acceptance and Rejection of Evolution | 59 |
| 4.1 | Cross-National Acceptance of Evolution | 91 |
| 4.2 | Statistical Relationships for Secular, Compromise, and Creationist Survey Responses | 97 |
| 4.3 | Intercorrelations of Opinions Supporting and Rejecting Evolution in Texas | 100 |
| 4.4 | Movements in Opinion Concerning the Origin and Development of Life on Earth | 108 |
| 4.5 | Movements in Opinion Concerning the Evolution of Human Beings from Animals | 109 |
| 4.6 | Movements in Opinion Concerning the Origin and Development of Human Beings | 110 |
| 4.7 | Difference-in-Differences Analysis | 114 |

# ACKNOWLEDGMENTS

Only those who have written a book can understand how much the existence of a volume that has a single name on the cover nevertheless would not exist but for the crucial assistance of many other people. In this case, I asked various scholars in special fields to read and critique individual chapters. They made many sensible suggestions. Naturally I did not implement every suggestion; still, this book is much different, and much better, because of their efforts. I fear I will never be able to adequately compensate them. I can only list them, and hope that the listing will bring them public acclaim:

## Chapter Readers

1. Benjamin Gregg, Professor of Social and Political Theory, University of Texas at Austin
2. Robert Sprinkle, Associate Professor, School of Public Policy, University of Maryland
3. Patricia Kelley, Professor of Geology, University of North Carolina at Wilmington
4. Brian Roberts, Professor of Government, and Tse-min Lin, Associate Professor of Government, University of Texas at Austin
5. Lino Graglia, A.W. Walker Centennial Chair in Law, University of Texas at Austin
6. Daniel P. Franklin, Associate Professor of Political Science, Georgia State University

I have to give special credit to my colleagues Bryan Roberts and Tse-min Lin. Their mathematical knowledge greatly exceeds my own. They worked with me on two projects that eventually evolved into the material in Chapter Four.

## Others Who Helped in Various Capacities

Dan Bolnick
Jay Budziszewski
Arturo de Lozanne
Gary Freeman
Gary Jacobsohn
Bryan Jones
Robert Koons
Richard Lewontin
Eda Matthews
Nancy Moses
Lorraine Pangle
Thomas Pangle
Brian Roberts
Marie Roberts
Daron Shaw
Neil Shubin

# INTRODUCTION

*The Politics of Evolution*

In the summer of 2000 I was flying back to Austin from France, and I had a ten-hour layover in O'Hare Airport in Chicago. Desperate for something to read to help pass the time, I wandered into the little bookstore in my terminal. Most of the books in that store—romance novels, spy thrillers, and self-help manuals—did not appeal to me. But on the non-fiction shelf I noticed a copy of *The Lying Stones of Marrakech*, a collection of essays by Harvard paleontologist Stephen Jay Gould.[1] I had read Gould's first collection of popular essays, *Ever Since Darwin*, back in the 1970s.[2] I remembered nothing about it, except that it was a good read. So I bought this new offering, and got about halfway through it before it was time to board the plane.

Reading the new collection I was struck, as I had not been upon encountering Gould for the first time, by the fact that his mind seemed to work along two tracks at once, the scientific and the political. Every essay was about some interesting fact of natural history—the controversy between "gradualists" and "catastrophists" among nineteenth-century geologists, say, or the way that biologists finally found evidence for bacterial life more than 2 billion years old by looking in chert beds rather than regular sedimentary rock layers. But each essay also contained a political moral, which was itself often wrapped in a lesson about scientific method. I had been reading popular science, off and on, for decades, but I had never encountered in that genre a mind that seemed so in tune with my own, alert to the political implications of every act and fact.

When I got back to work in Austin, in my spare time I began buying and reading more of Gould's books. It soon became clear that he was embroiled in a series of controversies within the profession of evolutionary biology. Many of his essays were clearly responses to something other biologists had done or written. Some of his efforts seemed to be intended to refute arguments made by others who had attacked his own ideas. And all of these controversies seemed to be both political and scientific at the same time.

Hooked, I began to read Gould's critics. It became clear that the world of evolutionary biology was the battlefield for two types of wars, one, "outside," between the profession as a whole and resurgent creationism, and one, "inside," between various ideological factions within the profession. The two types of conflict sometimes interacted and were powerfully influenced both by methodological issues and by disagreements over the philosophy of science. There was politics everywhere.

I decided that I wanted to study this subject in a serious way. I joined the Society for the Study of Evolution, read its journal, and attended one of its conventions. I audited a graduate course in "Speciation" given by Professor Dan Bolnick, with his kind permission, of the Department of Integrative Biology at the University of Texas. I sat in on technical talks by members of the various life-sciences faculties at UT. I checked out classic works in the field and read them with a *Dictionary of Biology* at my elbow to help me master the vocabulary.

After all of this I wrote *Stephen Jay Gould and the Politics of Evolution*, published by Prometheus Books in 2009. I was reasonably satisfied with that book, and the reviewers generally treated it kindly. But still, I felt that its narrow focus had caused me to expend most of my time exploring "inside" politics. Most of the substance was about arguments between specialists, much of it over technical issues. The technical issues all had interesting political implications, but still, I felt that I had not covered enough of what was important in the social conflict over evolution. So, I decided to write another book, one whose focus would be mostly on "outside" politics. In this second book, incidentally, there would be almost no mention of Stephen Jay Gould.

The conflict between almost all the members of the profession of evolutionary biology, in alliance with educators, judges, and many members of the concerned public, on one side, versus creationists and their allies on the other, has very widespread implications for society as a whole and for the activity of science in particular. In this book, I discuss those far-reaching implications. But in a specific, practical sense, the arena in which the conflict is usually fought out is the public schools. What is to be taught in public school biology classes? This more focused question, because it is the flashpoint of the politics of evolution, will occupy more of my pages than any other topic.

I try to write in clear English as much as possible. But there are a few technical concepts that require explanation, both because their meaning is highly abstract and because they are not always used the same way by everyone. Here, then, is a short discussion about how I will use some important terms.

"Ontology" refers to the broadest possible assumptions of the human mind. How are we to characterize the fundamental forces at work in the universe? Is there such a thing as an intelligent spiritual power that must be understood as the organizing principle of all matter, energy, and life? Or are the powers that mold the universe physical and unconscious? Are the forces at work in the universe the same at all times and places, or might they have been different at some point in the past, and in some place far away? What is the nature of causation, and are there natural laws that govern its operation? In the pages of this book I will often have occasion to refer to the different ontological assumptions of various players in the political dramas. Especially, creationists argue from different ontological assumptions than do scientists. Thus, a specific argument over a government policy often can be traced back to the largest, most general assumptions possible about the nature of reality.

The different ontological assumptions make for different evaluations of the meaning of the word "secular." It means non-religious. It does not mean anti-religious, although a secular person can also be anti-religious. The outlook and methodological stance of science is secular. Scientists are not anti-religious—many scientists, now and in the past, have been quite devout. But all science is secular in its methodology. Everything must be empirically grounded, and thus measurable. So scientists, religious or not, must avoid religion in discussing the evidence for their hypotheses.

To people with a religious ontology, however, secularism is impossible. If all matter and energy is suffused with spiritual purpose, then there can be no non-spiritual methodology. There can be no science without an acknowledgement of the reality of the miraculous. To people who hold to a certain type of spiritual ontology, therefore, secular science is anti-religious, anti-reality, anti-God. It is not only bad science; it is wicked.

As a consequence, when many people in our culture argue that the public schools must be secular, they believe that they are being even-handed and fair to all religious sensibilities. But to many creationists, such an attitude is both nonsensical and the advocacy of spiritual tyranny. There is the right way to teach children and the evil way, and secularism is just one variation on evil.

Finally, the conflict is over a scientific theory that, in its modern form, has achieved the endorsement of the overwhelming majority of practitioners in the life sciences. As propounded in essays by Charles Darwin and Alfred Russel Wallace in 1858, and then in more detail by Darwin in *The Origin of Species* in 1859, it seeks to explain the development of the profusion of life-forms on planet Earth. To this theory was added, in the twentieth century, an understanding of genetics, first theorized by the Austrian monk Gregor Mendel in the 1860s. By about 1940, biologists had melded the two strands into the Modern Synthesis. In common political discourse, when a writer uses the term "Darwinism," it is often not clear whether the term refers to the original 1858 formulation by Darwin and Wallace or to the luxuriant development of the science of evolution after the emergence of the Modern Synthesis, or, indeed, whether the writer understands that biology has developed mightily since Darwin. That is one reason why biologists tend to avoid the term "Darwinism."

Understanding the theory requires an effort of mental readjustment, for Darwin's definition of "species" is different from the unspoken assumption that seems obvious to the untutored human mind. When it is observed at a specific historical moment, a species, to most modern biologists, is a definite entity, a group of animals, plants, sponges, bacteria, etc., the members of which share specific characteristics. But when it is viewed in the fullness of history, a species is simply the outward manifestation of a present arrangement of genes, an arrangement that changes constantly through deep time. Our minds are fooled into thinking that there is something essential about every species—the "kinds" mentioned in the *Bible*. Because of our short life spans, we do not understand that when we look at, for example, a cat, we are seeing a single snapshot of an infinitely varying array. A million years ago, the ancestors of this cat were constructed by a different set of genes and looked startlingly different from the feline in front of us. A million years hence, the successors to this outward manifestation of an inner collection of genes will be different again. But there is no specific instant in the historical sweep when a cat is born from a non-cat. Viewed from the perspective of geological time, there is no essential cat; no kind.

At any given moment, runs the theory, the organism that represents that moment's collection of genes is confronting an environment full of danger and opportunity. Some of the organisms will possess characteristics that make them superior to others that are competing for the same resources. They will be able to withstand higher temperatures, or they will be able to run faster, or they will be more intelligent, or they will have neural characteristics that allow them to cooperate. They will thus survive, breed, and pass on their genes. Their competitive adversaries, possessing traits with lower survival value, will die without breeding, and their genes will be lost. Over deep time, the organisms representing the gene combinations conferring superior survival and breeding skills will pass on those combinations and thus their changing forms and behaviors, and "new" species of organisms will appear in abundance. Darwin termed this theory of the mechanism that causes organismal change over time "natural selection." The long-term result of that mechanism is the process that history has called evolution.

Modern evolutionary theory identifies other mechanisms for change in gene frequency within populations, including gene flow and genetic drift.[3] So, to scientists, natural selection is only one cause of evolution. In the sloppiness of public discussion about "Darwinism," however, these distinctions tend to be lost. In this book, therefore, I will often be discussing somebody's ideas about "Darwinism" that make no sense. But senselessness is part of the political struggle. I will therefore often fail to draw distinctions (such as between natural selection as a means and evolution as an end result) that are important to scientists but have no traction in political argument.

A further reason for the ineradicable ambiguity of public discussion is that even among biologists, there is a variety of opinions about the details of the science of evolution. (For example, in their textbook, Coyne and Orr report that there are at least twenty-five ways to define the concept "species."[4]) As a consequence, my summary of the theory would not be endorsed by everyone in every branch of the life sciences. Indeed, biologists as a group are cantankerous, fiercely individualistic, and quarrelsome—and I say that with love. Every specialist has a slightly different take on what natural selection (as the most important cause), and evolution (as the end result) mean in their details. In my characterizations of the theory in this introduction and in the rest of the manuscript, I have tried to capture the central tendencies of thought in the life sciences. But there is probably not a single biologist in the world who will agree with all of my assertions. And don't get me started on the philosophers of science. In my book on Stephen Jay Gould's ideas, I explicated and evaluated many of the controversies that are a permanent feature of the science of evolutionary biology. I will not go over that ground again in this book. I must ask life scientists, and others with expertise in this area, to refer to that book if they want to read my analyses of disputes among scientists and philosophers on specific issues, and to cut me some slack in this one.

Two corollaries to the theory are important. The first is that it does not claim to explain how the first organism (which presumably was a self-replicating molecule, but nobody knows) came about. The theory of natural selection asserts that it can explain everything that came after that molecule. Creationists like to emphasize the inability of evolutionary biologists (so far) to explain the origin of life, as though those scientists were keeping it a secret. But no biologist, from Darwin forward, has denied that the origin of life is an unsolved scientific puzzle.

A second point about the theory is that, in broad outlines, it has been supported by an overwhelming amount of evidence accumulated over more than a century and a half. It has performed particularly impressively in the toughest area of scientific method—prediction. People (even scientists) sometimes say that natural selection is not a predictive theory. That is a true statement only if the idea of prediction refers to something that will happen in the future. The theory of natural selection cannot, however, predict the future evolution of a species, because that evolution will be the result of the interaction of the species with its environment, and no scientist can predict the nature of the environment that the individuals in any given species will encounter.

Nevertheless, evolutionary biology, in its many specialties, has been hugely successful in predicting things in the present and the past. In every branch of the life sciences, scientists have deduced predictive hypotheses from the general theory, gone to look for evidence that would contradict or support the theory, and found supportive data.

Here is one out of a very large number of examples that could be offered. In the early 2000s, a professor of anatomy at the University of Chicago named Neil Shubin was thinking about what the theory of natural selection predicted about the transition between fish and amphibians. No evidence had been found in the fossil record of a creature that was a transitional species between water- and land-dwelling organisms. (Creationists had been using this hole in the history of the rocks as evidence for their own claim that the fossil record contradicted the theory.[5]) But that lack of a transitional fossil was bracketed by solid evidence. According to the paleontological record, in rocks from about 385 million years ago, there are only fish fossils; in rocks from about 365 million years ago, there are fossils of fish, reptiles, and amphibians.

Shubin reasoned that there must be evidence of a transitional creature in rocks of about 375 million years old. That is, he made a prediction. He got a grant, found an accessible area of the Earth on which such rocks were to be found—Ellesmere Island in the Canadian arctic—assembled a team, and went fossil-hunting. After four summers spent sleeping in tents, avoiding polar bears, and hammering away at 375-million-year-old rock layers, the team found a fossil of an organism that is an almost perfect transition species between fish and amphibians. This creature, to which they gave the Inuit name *Tiktaalik*, is a blend of fishlike and amphibian-like characteristics:

> Like a fish, it has . . . fins with webbing. But, like early land-living animals, it has a flat head and neck. And, when we look inside the fin, we see bones that correspond to the upper arm, the fore-arm, even parts of the wrist. The joints are there, too; this is a fish with shoulder, elbow, and wrist joints.[6]

It is easy to focus on the exciting fact that this discovery adds significantly to the sum total of human knowledge about the past and lose sight of the even more important fact that the discovery was the result of a *prediction*. The theory of natural selection gave rise to the prediction, and the empirical results of the prediction created greater confidence in the theory.

Given that the theory has passed thousands of such predictive trials over a century and a half, it might be assumed that it would long ago have become accepted as the truest explanation we have of the history of life. Within science, it has. But in the wider society, it labors under a permanent cloud of suspicion. That cloud has political consequences.

Why that suspicion exists, how it is expressed in political attitudes, how those attitudes are translated into rhetoric, and the ways that they have worked their way into American party politics and Constitutional law are, partially, the subjects of this book. Partially, the subject is how the pro-evolution people have responded to doubts about the theory. The book is about the controversy as a whole. But because so much of the controversy in the United States has found expression in a conflict over the content of the public school biology curriculum, much of my specific attention will be focused on that issue.

## Notes

1. Gould, Stephen Jay, *The Lying Stones of Marrakech: Penultimate Reflections in Natural History* (New York: Harmony Books, 2000).
2. Gould, Stephen Jay, *Ever since Darwin: Reflections in Natural History* (New York: W.W. Norton, 1977).
3. Readers interested in the various mechanisms that figure in modern theories of evolution may want to acquire Jablonka, Eva and Marion J. Lamb, *Evolution in Four Dimensions: Genetic, Epigenetic, Behavioral, and Symbolic Variation in the History of Life* (Cambridge, Mass.: MIT Press, 2005).
4. Coyne, Jerry A. and H. Allen Orr, *Speciation* (Sunderland, Mass.: Sinauer Associates, 2004), 25.
5. Johnson, Phillip E., *Darwin on Trial*, 2nd ed. (Downers Grove, Ill.: Inter-Varsity Press, 1993), 76–77; Yahya, Harun, *Fascism: The Bloody Ideology of Darwinism* (Istanbul: Arastirma Publishing, 2002), 228–229.
6. Shubin, Neil, *Your Inner Fish: A Journey into the 3.5 Billion-Year History of the Human Body* (New York: Random House, 2009), 23.

# 1

# BIOLOGY

## The Most Political Science

On the "Public Affairs" shelves at Barnes and Noble, and sundry independent book stores, there are many lengthy discussions of Problems. The same phenomenon occurs in cyberspace in a less tangible setting—an endless number of websites and blogs devoted to conveying a certain interpretation of a Problem. The Problems are numerous and serious: global warming, the moral decline of America, a growing economic inequality, the menace of unrestricted immigration, worsening pollution of rivers, Islamic terrorism, acidification of the oceans, evil liberalism, evil conservatism, the inexplicable popularity of reality television, and dozens more.

Although I have not read every book or perused every Problem site on the Internet, I have sampled enough of them to be able to make a generalization. The authors who are warning about the Problem all have a Solution. The Solutions differ. Sometimes the Solution is a return to free-market capitalism. Sometimes it is to cherish and nurture the Earth. Sometimes it is to return to God. Sometimes it is to make our government so small it can be drowned in the bathtub. Sometimes it is to make our government so powerful no outsider would dare to menace us. Sometimes it is to raise taxes on the rich. Sometimes it is to make it harder for the poor to vote. Sometimes it is to junk our present Constitution and write a new one. There are as many Solutions as there are Problems.

This book is different, at least partially. Its subject is indeed a Problem, although the Problem has several facets. The Problem is the many philosophical, jurisprudential, and political issues raised by the dominance of the modern theory of evolution in biological science. The theory, which was originally articulated in 1858 by Charles Darwin and Alfred Russel Wallace, rather quickly became a dominant paradigm in the life sciences, and even more so after it was merged with genetics during the twentieth century. But because of its implications for religious and, often, non-religious philosophy, it has always spawned fervent opposition from some citizens. Over the last few decades of the twentieth century and into the twenty-first, that opposition has found political expression. The political conflict is most vivid and widespread in regard to the question of whether the scientific theory of natural selection (as a cause), and the process of evolution (as a result), should be taught exclusively in public school biology classes, or whether it should be replaced, or accompanied, by some version of creationism. So "Darwinism," as it is usually labeled (although not by biologists) is a Problem. Whether the Problem is that it exists, or that it is being attacked, is part of the Problem.

The difference between this book and all those others, however, is that I do not have a Solution. Or, more technically, I have several tentative solutions, and they tend to be inconsistent with one another. My own thoughts in regard to the politics of evolution, therefore, tend to be ambiguous, and my feelings ambivalent. In regard to this inner tension, as I will explain in some detail, I represent the United States citizenry in microcosm. Evolution presents us with a situation that political scientist Deborah Stone terms a "policy paradox"—in which each potential choice violates a value we hold dear.[1] But that fact makes it an interesting subject to explore.

This first chapter will mostly be devoted to examining a conflict within the heart of scientific investigation—the extent to which science is, and should be seen as, independent of politics. The subject of the second chapter is the rhetorical side of the controversy between scientific biology and its enemies, and, in particular, the use of metaphor in selling and attacking it. The subsequent four chapters are all addressed to various aspects of the political controversy within the United States. In Chapter Three, I explore the relationship between evolutionary biology and religion, and apply my conclusions to American politics. In Chapter Four, I look at public opinion in regard to evolution and the policy question of what should be taught in public schools. In Chapter Five, I examine the jurisprudence surrounding the policy question. In Chapter Six, I look at the way the political process has dealt with the issue, thus creating a democratic politics that contradicts its own professed values.

## The Politics of Implication

Most scientists live with an ideological tension. On the one hand, they want to pursue the truth about nature in a neutral, disinterested manner. In addition to the common human disinclination to be bothered by authorities, curious dunces, pious busybodies, and tattletales, the scientific personality is similar to the artistic personality in being motivated by a ferocious inner need to *get it right*, and therefore to resent intrusions from outside.[2] The scientific is different from the artistic character in its orientation to empirical reality rather than inner visions, but its resentment and avoidance of outside interference is similar.[3]

There are, in addition, good and oft-explained reasons for scientists to be permitted to go about their business unmolested. Not only scientists themselves, but anyone who understands the benefits of free thought can respect the desire to be unrestrained in the pursuit of truth. The scientific preference to be free of politics is itself a political stance, of course, but it is a stance that has been justified by the success of scientific activity over several hundred years.[4]

From the initial glimmerings of the modern scientific attitude in Europe in the sixteenth century, scientists worried about the potential nuisance of outside interference. The first philosopher of science, Francis Bacon, fretted about the "troublesome adversary" of science, the "blind and immoderate zeal of religion," which has, through the "simpleness and incautious zeal of certain persons been drawn against freedom of inquiry."[5] Bacon's strategy for meeting the threat was to firmly and frequently assert that true science had no philosophical pretenses, and that its only motivation was to bring the "gift of reason to the benefit and use of men."[6]

When the first scientific organization, Britain's Royal Society, was founded in 1662, its members collectively adopted the same strategy of avoiding the "passions, and madness of that dismal Age," according to Thomas Sprat, its first historian, by forbidding discussion of religion or politics at its meetings. Only the "News" of "Physick, Anatomy, Geometry, Astronomy, Navigation, Statics, Mechanics and Natural Experiments" were appropriate subjects for scientific conversation, and the exclusive purpose of the membership was to "judg[e] and resolve upon the matter of Fact."[7]

The insistence of scientists on their purity and independence from social conflicts and non-scientific authority, in other words, was part of the scientific self-concept from the beginning of the enterprise. And, given the spectacular achievements that such independent activities have brought to human life, the non-religious, not-political strategy would seem to have been endorsed by history.

That is certainly the position of most scientists, and some philosophers of science, in contemporary society. As philosopher Abraham Kaplan has put it, the only conditions under which science can exist are "freedom of inquiry, of thought, and of its expression," which, of course, sum to independence from outside interference.[8] Biologist Kenneth Miller speaks for many other natural scientists when he states, "For me, one of the great attractions of science has always been its profoundly nonpolitical nature."[9] And legions of scientists have warned their colleagues to guard against the "corruption of science by politics."[10] Some are so ardent in their insistence upon their own independence that they would actually claim that they have a *right* to do scientific research.[11]

The theoretical separation of scientific activity from practical relevance was established on a philosophical foundation by David Hume in the eighteenth century. There is no logical or empirical justification, wrote the Scots philosopher, for the attempt to deduce a moral precept from an empirical fact. Facts exist alone, independent of moral theories. Whatever scientists decide about the reality of the world, it has no bearing on how humans ought to behave. Ever since, the great majority of scientists have argued that people who were trying to derive moral rules from the facts of nature were breaking "Hume's law," committing the "naturalistic fallacy," and thus disqualifying themselves from scientific discourse.[12]

After Hume, most scientists who addressed the issue repeated the point. "[A]nyone engaged in research or in presenting its results should keep two things absolutely separate," wrote the influential and oft-quoted German sociologist Max Weber, "first, the statement of empirical facts . . . and secondly, his own practical value-position."[13] And more to the point of the subject of this book, Rosenberg and McShea, in *Philosophy of Biology*, state the rule clearly and succinctly: "[T]he theory of natural selection has no normative components, hidden or obvious, implicit or explicit."[14]

It is easy to see the political value of this philosophical position. In the twenty-first as in the seventeenth century, scientists in general, and biologists in particular, will not have to defend their own project against charges that its consequences are morally dubious if they can convince the public that science has no moral consequences. There is no need to defend that which does not exist. Whether Hume's law is true or mistaken, therefore, it is greatly useful to scientists who want to be left alone.

But there is another hand. For a variety of reasons, scientific activity cannot simply be divorced from the interests and worldviews of the outside society. In the first place, much modern scientific research is funded, directly or indirectly, by tax money. In order to achieve the "holy grail of money without interference," in Daniel Greenberg's words, individual scientists and scientific organizations have to play the democratic political games of interest-group politics, lobbying, and campaigns to influence public opinion.[15]

In the second place, while freedom of inquiry is a lovely concept, it ignores the fact that the direct results of scientific investigation can be horrendous. From Mary Shelley's novel *Frankenstein* in 1818 to a spate of motion pictures in the twentieth and twenty-first centuries (*Forbidden Planet* in 1956, *Jurassic Park* in 1993, and *Deep Blue Sea* in 1999 are three examples), fiction writers have imagined the terrible consequences that can accompany scientific understanding. The fiction works because it seems to apply plausibly to a possible non-fiction future. Moreover, the well-publicized regrets experienced by physicist Robert Oppenheimer when realizing the destruction he had helped cause by supervising the Manhattan Project illustrate the non-fictional horrors that can be the result of scientific activity.[16]

In the third place, many philosophers, sociologists, and historians of science, and some scientists themselves, have argued that scientific activity is only possible in a specific historical and societal setting, and that the assumptions and dominant ideologies of that setting are unalienable ingredients of the thinking of scientists. In other words, whatever scientists would prefer, politics is an inescapable part of their profession. As geneticist Richard Lewontin and ecologist Richard Levins put it, "scientists, whether they realize it or not, always choose sides."[17] This particular view is contested among scholars, and it is by no means settled that the dominant cultural assumptions of any given era color the empirical findings of that era. Nevertheless, it is not a good idea to assert the independence of scientific inquiry without at least a nod to the thinkers who maintain that there is no such thing.

These three types of the mixing of the scientific enterprise with real-life interests and ideologies have been studied at some length, and while acknowledging them here, I do not intend to pursue them. My interest is in a fourth type of real-life consequence of scientific advance: Scientific theories and discoveries have *implications* for the intellectual world outside the laboratory, and those implications have political consequences. Or, to put the point more finely, whether or not science has social consequences, many people believe that it does, and that belief serves as the motive force for a variety of political controversies.

Prior to the seventeenth century, "science" did not exist as an independent mode of inquiry. In European as in other cultures, the activity of investigating nature was indistinguishable from the faith-system of Christianity. Religion, philosophy, and science were of a whole. As historian H. Floris Cohen expresses the idea, "knowledge of nature was held to make sense only in the framework of a comprehensive view of the world and man and how these are connected."[18] But the process of scientification that began to shake European society in the seventeenth century was, as illustrated by the self-restraint of the members of the Royal Society, a process of withdrawing from the philosophical context of the outside world. Not only in England, but in Italy, as with Galileo, and in France and the Netherlands, as with Descartes, there arose a "new science" that "claimed to provide an understanding of reality that did not owe its warrant to whether or not it fitted in with an all-encompassing insight into the order of the world but only to the extent to which it satisfied inherent criteria of a methodological nature."[19] Scientists tried to buy immunity from supervision by claiming that their interests and goals were so limited as to not form a challenge to any non-scientific understanding of the cosmos.

Nevertheless, by the time it had become a social force, in the nineteenth century, the scientific worldview had achieved the status of an "all-encompassing insight" of its own, one that was radically at variance with its parent ideology. A major social activity that eschews moral standards, yet awards its practitioners social power, is certain to offend a significant proportion of its audience. Whereas many scientists have agreed with Kenneth Miller that the "nonpolitical" nature of their career is one of its most attractive aspects, they cannot escape the fact that the essence of their activities challenges the values and perspectives of millions of people in their own society.

Concepts such as "worldview," "all-encompassing insight," and "ideology" are extremely difficult to define, and even harder to measure. In this book, therefore, I must treat one of my chief concepts, *implication*, with an unsatisfactory imprecision.

In scientific discourse, "implication" can have an unambiguous definition—the observable consequences that must follow if a theory is true. The implications of Einstein's general theory of relativity, for example, were that a large mass, such as the sun, would bend light rays so that starlight passing close to the object would be deflected, making those stars appear to be in different positions than they appeared to occupy at other times. In this usage, a hypothesis is an implication in its operationalized form. The hypothesis derived from the theory could be tested against empirical reality, and, in the case of Einstein's, it was. The hypothesis having been borne out, the theory was provisionally accepted by the scientific community.

The formal scientific meaning is not the way I use the term "implication." This is a book about politics. Politics is partly about the way people interpret the physical and moral reality they see around them. Empirical evidence, so important in science, is of much less importance in political argument. In politics, implication is an interpretation to be advocated, and therefore, something to be fought over. Because there is a politics of science, there are interpretive struggles over the implications of given scientific doctrines.

In politics, an implication is a consequence that, to some observers, would seem to follow inevitably from some other action or belief, without actually being necessitated by logic or physical connection. It is the ideological by-product, the mental side effect, the stowaway on the voyage of discovery, the fellow traveler of a discourse, the theory's sidekick. An assignment of implication is, in general, a large, vague, complicated causal theory, in which one set of ideas and actions in one area of human affairs is accused of contributing to the vitality of another set of ideas and actions well separated in time and space. Was the Protestant Reformation implied by the invention of the printing press during the fifteenth century? Did the accession of Protestantism in parts of northern Europe during the sixteenth century imply the rise of capitalism? These questions cannot be answered definitely, although many scholars have tried. Did Darwin in the nineteenth century imply Hitler in the twentieth? Although many scientists, and pro-science citizens, would reject that question as absurd on its face, a significant percentage of the population would answer it in the affirmative.[20]

Implication is not to be confused with "externalities," a concept from economics that conceives of the costs of an activity that are not directly measured by the ordinary methods of calculation. If a factory pours its effluent into a river, for example, it is shifting the costs of cleanup to the people who live downstream, or to the environment. This connection differs from implication in that it can be observed and measured. Pollution is thus an externality, not an implication.

Nor is an implication the same as an "unanticipated consequence," a concept long used by political scientists to caution policymakers about the hazards of making changes in society. The problem of trying to imagine all the possible consequences of an action is intense and unresolvable, for there are potentially a very large number of ways that human action can change the world, and most of those potential effects are beyond imagining. Global warming, for example, is an unanticipated consequence of industrialization. But again, the connection is in retrospect observable and measurable, and generally exists in the realm of physicality and action rather than beliefs and values.

Nor is an implication the same as "emergence," a phenomenon that occurs when two or more things come together to make a third thing that has properties that were not predictable from knowledge of the properties of the two constituents. For example, when sodium, a poisonous metal, and chlorine, a poisonous gas, are combined, the result is sodium chloride—table salt—which has many qualities not found in its two constituents, including human edibility. Emergence is, in fact, an oft-used concept in evolutionary biology. Scholars argue about whether the upper categories of life—families or orders, for example—have qualities that cannot be deduced from the qualities possessed by their constituent species. But this is not the same sort of argument as the ones that swirl around implication.

Debates about implication, in general, tend to have a background in very large and important issues of historical causation, but, in specifics, they often comprise arguments about esoteric and obscure details. Frequently, scientific controversies are fought out within the realm of methodology, with each alternative technique somehow implying different social consequences. Although the contestants in a scientific scuffle often fail to mention, or actively deny, the political implications of the methods they employ, outsiders often have no trouble understanding the big issues at stake in the tiny arena.

The history of science is thus full of fights over implications, sometimes centered on the actual theory, sometimes on the empirical findings, sometimes on the methodologies employed, and sometimes on all three together. There are many examples of the way that assumed political implications underlie even some very technical aspects of these controversies. There are numerous examples:

(1) *The heliocentric theory.* This is the granddaddy of all struggles over implication. Although its unofficial practice was not always consistent with its official policy, for a millennium and a half the Roman Catholic Church had officially accepted as physical truth the apparent endorsement in the *Bible* of the belief that the Sun orbited the Earth. When Galileo began to publicize his own view that Copernicus' 1543 theory that the Earth orbited the Sun was correct, the Church objected. In 1633, the Inquisition forced Galileo to recant, under threat of execution.

Modern historians have argued that the Church's objection to Galileo's writings were not so much that they contradicted its teachings, but that he presented them in such an arrogant and dogmatic style that he directly challenged the Church's ideological authority.[21] I find this argument to be irrelevant even if it is correct. If Galileo had presented arguments *supporting* the Church's teachings in an equally arrogant and dogmatic manner, he would not have been molested; indeed, he probably would have been rewarded. Content was crucial; style was pretext. Galileo was persecuted because of the "ethical (social) implications" of his doctrine, according to philosopher Paul Feyerabend. "Heresy, defined in a wide sense, meant a deviation from actions, attitudes and ideas that guarantee a well-rounded and sanctified life." The Inquisition stepped in because it "wanted to protect people from being corrupted by a narrow ideology that might work in restricted domains but was incapable of sustaining a harmonious life."[22] In other words, Galileo was almost burned at the stake for an implication.

(2) *Hobbes, Boyle, and the air-pump*. In the 1660s and 1670s, the great political philosopher Thomas Hobbes and the great chemist (and, not incidentally, one of the founders of the Royal Society) Robert Boyle engaged in a rancorous public debate about the efficacy of a piece of technology. Boyle had invented, or at least greatly improved, a device for pumping the air out of a glass globe, thus creating a vacuum. He used the air-pump to engage in a variety of experimental inquiries into scientific questions, the details of which need not concern us. Hobbes attacked his experiments publicly, making specific criticisms of the pump (charging that it did not work—did not actually create a vacuum) and, more importantly, attacking the assumptions underlying Boyle's whole project.

In a number of writings, Boyle had articulated a protocol for conducting scientific experiments—how they were to be designed, how hypotheses were to be tested, how the members of the community of scientists were to evaluate the conclusions thus reached, and so on. Hobbes, whose own devotion to absolute monarchy had been thoroughly set forth in his political-theory masterpiece *Leviathan* in 1651,[23] realized that in shifting the imprimatur of truth from the king, applying right philosophy, to the Royal Society, evaluating experiments, Boyle was setting up an authoritative institution that was independent of the sovereign.

As he had made memorably clear in *Leviathan*, Hobbes's position was that only a single, absolute authority could prevent the sort of political controversies that led to a breakdown of civil comity, followed swiftly by the arrival of murderous chaos. Thus, any plan for making the king's authority less than universal in application was, by implication, a step along the road to anarchy.

As the historians Shapin and Schaffer put it in their account of the air-pump controversy, "Hobbes said that no independent group of intellectuals could avoid constituting a threat to civil society."[24] Empiricism—Boyle's experimental method—attacked social order, because the evaluation of empirical results required a group of experts separate from the sovereign. Any such clique inevitably bred political dissent, thus paving the way to civil war. "For Hobbes there was no philosophical space within which dissent was safe or permissible.... [T]he rejection of the vacuum was the elimination of a space within which dissension could take place."[25] The implications of what has come to be accepted as scientific method were inconsistent with social peace, and thus hostile to human life.

While on the surface, the controversy seemed to be about the specific issue of the air-pump, therefore, of far greater importance was the subterranean and general issue of philosophy. To Hobbes, political philosophy and philosophy of science were identical. To Boyle, they were distinctly separate.

Boyle won the argument in the seventeenth century. But Hobbes's insistence that scientific methodologies inevitably have metaphysical and political consequences is an idea that has not gone away.

(3) *The Big Bang*. In the 1920s, astronomers discovered that the universe was expanding; that is, that the galaxies were all rushing away from each other. The awareness that it was getting bigger naturally spawned the opposite thought, that it must have been smaller in the past, and in fact at some time must have been just a point, from which it exploded into the cosmos we observe. After the initial insight, physicists, astronomers, cosmologists, and mathematicians applied calculations and measurements to the theory. By the 1970s the notion of a single origin moment had become "the Standard Model," in the words of physicist Steven Weinberg, "a theory that we take as our working hypothesis."[26] The great majority of specialists now endorse the idea that the universe began in a micro-moment of expansion about 13.7 billion years ago.

Not every natural scientist, however, has agreed with the consensus. In particular, Fred Hoyle, a prominent British cosmologist, was unconvinced, so much so that while being interviewed on a radio program, he referred to the theory in contemptuous dismissal as "the Big Bang."[27] In ironic reversal of Hoyle's intentions, the phrase caught on, and is now the shorthand label used by scientists and non-scientists alike, as witnessed by *The Big Bang Theory*, a popular television comedy program in the United States.

Hoyle had scientific reasons for resisting the consensus, but he was also motivated by a suspicion of the implications residing in the idea of a universe popping suddenly out of the void. An atheist, he disliked the encouragement that he expected the notion that "[i]n the beginning, there was an explosion"[28] would give to fundamentalist Christians, who would see in it a confirmation of the "let there be light" creation myth that begins the *Old Testament* of the *Bible*.

As it happens, the reception that Christians have given the Big Bang theory has not been uniform. In fact, Christian attitudes have been so diverse that they illustrate one of my contentions about the political implications of scientific theories: Because implications are malleable, and impossible to refute or confirm, they lend themselves to contradictory rhetorical purposes.

In 1951, the Catholic Church, possibly feeling burned by the historical fallout from its treatment of Galileo, and wanting to get ahead of the scientific curve on the next great cosmological insight, officially accepted the Big Bang as proof of *creation ex nihilo* (creation out of nothing) and the existence of God.[29] Some

Protestants have also cited the Big Bang as evidence that "Science, done right, points toward God."[30] Among other prominent theists who write about the implications of science, William Dembski argues that modern thinking about causation, especially his own, compels the conclusion that "the cosmological theory of the Big Bang and the Christian doctrine of divine creation can now be brought into a relation of mutual epistemic support."[31]

But there are other Christian positions on the implications of the Big Bang. If one accepts the Standard Model as true, then one has also to concede that the universe, the solar system, and life on Earth are billions of years old. Further, one must endorse the methodology and technology of all modern natural sciences. "Young Earth Creationists," who believe that every word of the *Bible* is literally true and that therefore the whole shebang was willed into existence in 6 days about 6,000 years ago, cannot make such a concession. They must reject not only the conclusions of modern natural science, but all of its methodology.

Walt Brown, for example, who holds a Ph.D. in mechanical engineering from MIT, has put a great deal of time, effort, and thought into arguing that all modern physics, astronomy, chemistry, geology, hydrology, and biology—not just their theories, but their techniques of measurement—are completely mistaken. In Brown's new cosmology, the Big Bang never happened, at least in the sense that a natural event billions of years ago was the origin of everything we see and suspect. And he does not try to hide the larger truth that this smaller truth is pointing toward: "If the big bang is discarded, only one credible explanation remains for the origin of the universe and everything in it," and that explanation is to be found, of course, in a literal interpretation of the book of *Genesis*.[32]

Moreover, to some conservative Christians, it is not just that the factual assertions of modern science must be discarded if the *Bible* is to retake its place as the explanation of all being. Science itself is the enemy, because the purpose of science is progress in understanding; progress is by its very nature unstable; and instability is inimical to faith. Dr. Danny Faulkner, who seems to have learned a great deal of modern science, makes a coherent case as to why all that knowledge is to be disbelieved:

> A great concern of mine is that many Christians have wedded the creation account of the Bible to the big-bang theory, the current scientific myth of the world's creation. In a hundred years will anyone believe the big bang? If not, then what is to become of *Genesis* if we have tied it to the big bang? ...
>
> It is tempting to wed the Bible to our current understanding of the natural world, but that would be interpreting the perfect and unchangeable in light of the imperfect and changeable. Why would any Christian want to do that?[33]

The implications of the Big Bang theory, then, are contested as are the implications of many other theories in physics, chemistry, and astronomy. But these sciences together do not inspire as much theological dispute, political rancor, journalistic polemic, and judicial controversy as does biology, the science of life. Among its other intentions, biology seeks, directly or indirectly, to discover the facts of human nature, and as a result its basic project cannot escape being politically relevant. Further, because its major theoretical underpinning, the Darwin/Wallace theory of evolution by natural selection, unambiguously contradicts a literal interpretation of the creation story in *Genesis*, it cannot help but make enemies of a large proportion of the population of what used to be called Christendom.

(1) *Social Darwinism*. The most famous and execrated episode of the political application of biology began even before Darwin's *On the Origin of Species* appeared in 1859. In *Social Statics*, published in 1850, philosopher Herbert Spencer was feeling his way toward the argument that nature is on the side of the winners in society, and indifferent to the fate of the losers:

> Inconvenience, suffering, and death, are the penalties attached by nature to ignorance, as well as to incompetence—and also the means of remedying these.... [I]f there seems harshness in such ordinations, be sure it is apparent only and not real. Partly by weeding out those of lowest development, and partly by subverting those who remain to the never-ceasing discipline of experience, nature secures the growth of a race.[34]

When Darwin's book arrived nine years later, Spencer immediately latched on to it as a confirmation of his social philosophy.

The scientific theory of natural selection, as Darwin—and Wallace, although more briefly—explained it, is disarmingly simple. Because the environment is murderously ruthless, the organisms that have historically prevailed are those that have produced many more offspring than can possibly make it to adulthood, thus unintentionally ensuring a surplus of survival over mortality.[35] Further, although those offspring are recognizably similar to the parents, they may also differ slightly. Among the litters that are somewhat different from the parents, some individuals will possess traits that are more helpful for survival—their leaves are poisonous, they can run faster, they possess more effective camouflage, they are more intelligent, keener of eye, etc. These will be the offspring that are more likely to live longer and thereby have the time to reproduce, thus passing on those traits to their own progeny. Over immense amounts of time, the accumulation of beneficial traits in succeeding generations of better-adapted individuals, since the first replicating molecule, has resulted in the profusion of species that call the Earth home today.[36]

Neither Darwin nor Wallace understood how the variations that were a crucial part of their theory were generated or passed on to the next generation. It did not matter to Spencer. All that he needed was a general theory of how systems survived and changed through time. He had no need to look into the black box and see the system at work in its details. As it happened, at the beginning of the twentieth century, biologists rediscovered research done by Austrian monk Gregor Mendel into the particles of heredity, and named those particles "genes."[37] After several decades of argument, they combined the two theories—the macro explanation of Darwin and Wallace with the micro explanation of Mendel—into the "Modern Synthesis" that dominates biology today.[38] But for Social Darwinists of the nineteenth century, the macro explanation was enough.

To Spencer, the biological theory was merely a restatement of his own social theory about the progressive effects of merciless competition in the industrializing capitalist economies of Europe and North America. Spencer conflated economic success with "fitness" (a word that he, not Darwin, originated in 1864) and proclaimed in his post-Darwinian publications that he had discovered a scientific basis for laissez-faire economics.[39] Thus, Spencer assured the world that nature, as revealed by Darwin, endorsed monopolies, trusts, and the concentration of wealth and opposed welfare payments to the poor, public education, regulation of housing conditions, licensing of doctors, and even a state-supported postal service.[40]

As various historical studies have documented, what has come to be called Social Darwinism, but was really Spencerism, was intensely agreeable to the captains of industry of the late nineteenth century and was especially influential among the American judiciary.[41] In the United States, Spencer's influence was magnified by Yale sociologist William Graham Sumner, who translated Spencer's general message into a specific American context.[42] For several decades Spencerism dominated American social philosophy. It lost favor in the early twentieth century, however, as philosophers shredded its assumptions and economic hard times lowered the prestige of unregulated capitalism.[43] But its temporary prominence in pop intellectuality illustrates an important point about the relationship of biology and politics: Implications drawn from scientific discoveries and theories do not have to be accurate to be influential.

In fact, Spencer misunderstood the theory of evolution at a deep level. The only meaning that "success," or "fitness," could have in Darwinism was number and viability of offspring. Except for some vague rhetoric on the last page of the *Origin* about the way nature's competition forces each species to "tend to progress towards perfection," the theory of natural selection does not endorse the moral superiority of history's survivors. Still less does it give a biological endorsement to the human accumulation of wealth. Spencer's misapplication of evolutionary theory was a convenient rationalization for a specific social class during a brief historical moment. But that misapplication cannot be attributed to Darwin.

The historical impact of Spencerism was so vivid that many subsequent writers have assumed that "Social Darwinism" must always be politically conservative ("conservative" in the American sense of endorsing unregulated business competition and the resulting inequality of wealth and power, rather than in the European sense of defending a society dominated by monarchy, aristocracy, and an established church).[44] But it is not so. Any effort to deduce, infer, or draw religious or political implications from evolutionary theory is Social Darwinism, regardless of the values and conclusions of the thinker. And intellectual history has not lacked for non-conservative invocations of Darwinism. As Gertrude Himmelfarb summarized her historical survey of the uses of evolutionary theory in 1959:

> In the spectrum of opinion that went under the name of social Darwinism almost every variety of belief was included.... It was appealed to by nationalists as an argument for a strong state, and by the proponents of laissez-faire as an argument for a weak state.... Militarists found in it the sanction of war and conquest, while pacifists saw the power of physical force transmuted into the power of intellectual and moral persuasion.... Political theorists read it as an assertion of the need for inequality in the social order corresponding to the inequality of nature, or alternatively as an egalitarian tract.[45]

As is true of Darwinian theory in particular, so also is true of biology in general. The last two centuries have witnessed a long list of controversies over what, exactly, the science of life, properly interpreted, allows us to infer about the human condition. A few examples from this list will illustrate the variety of political fights that have been based on biological disagreements:

(2) *Nineteenth-century anthropology and the question of race.* Even before Darwin, scientists were offering differing answers to the question of whether all humans were members of the same species.[46] Humans differed in outward appearance—in height, skin shade, type of hair, width of nose and lips, and so forth—but did those differences merely divide people into subspecies (that is, races) or actually cleave them into a variety of species?

The touchstone for this question, as for all others at the time, was the *Bible*. At first, Christians believed that everyone was descended from Adam and Eve. Therefore, all humans must be variations on a single species. But as Europeans expanded their explorations of the world, often accompanied by political domination, they encountered many peoples whom they believed to be inferior to themselves. It would be much easier to justify the violence and domination that accompanied European cultural expansion if the subjugated peoples could be shown to be inferior, in some fundamental way, to their conquerors. It would be particularly soothing to the consciences of the members of the Caucasian subspecies if they could demonstrate that the races they dominated were actually animals rather than humans. The problem was especially acute for the grossest form of domination, slavery, and the most subjugated race, Africans and people of African descent.

Therefore, in the middle of the nineteenth century, European and American social theorists were divided on the question of whether all humans were in fact members of a single species (the "monogenist" theory) or whether the different races of humanity were actually different species ("polygenist"). When the theory of natural selection came along in 1859, it provided a way for people on both sides of the controversy to argue from naturalistic, rather than from Biblical, premises, and apply various kinds of empirical evidence. The dispute was so intense that two antagonistic anthropological organizations were founded in London during the 1860s. The Anthropological Society was for white-supremacist scientists who believed that they were of a different species than all the non-Europeans. The Ethnological Society (with Darwin as an honorary fellow) was home to those who embraced the monogenetic interpretation.

The polygenists had two major problems. First, their theory clearly contradicted the story of human origin in *Genesis*, which forced them to come up with some rather far-fetched interpretations of its words. Second, "species" and "race" were difficult words to define, leading to massive amounts of imprecision in supposedly scientific discourse. The standard definition of a species was of a group of animals or plants, the individuals of which could breed with each other and produce fertile offspring, but not with the individuals of other species. The standard definition of a race was of a subspecies, the individual members of which were recognizably different from the members of other subspecies, but who could nevertheless interbreed with those others. Because all humans, of all races, could demonstrably (and enthusiastically) interbreed and produce fertile children, polygenists had to originate imaginative new definitions for both species and race.

The problem for polygenists was most acute in the United States, where the biological controversy, prior to 1861, had direct, politically relevant implications for the slavery issue. Clearly, it would be far easier to justify the chattelization of blacks in the American South if they could be placed in a natural category that was inferior to the category inhabited by their owners. Pro-slavery anthropologists (not all of whom lived in the South) were so active in conducting research to prove that the African American slaves were of a different species from whites that polygeny was dubbed the "American school" by Europeans. The Philadelphia physician Samuel George Morton won wide notice, prior to his death in 1851, by reporting that, according to his measurements of typical skull sizes, the peoples of the world could be ranked according to the average cubic capacity of their brains. The brains of Europeans, he related, were largest, followed by Asians, American Indians, and, at the bottom, Africans. It took little effort to translate the size of brains into cognitive power, and thus, into a ranking of putative species by intellectual standing.[47]

The divisions on this issue were not clear-cut, because some prominent polygenists (notably Louis Agassiz) were anti-slavery, and because reliance on secular scientific research made many apologists for slavery, who were used to making their case by appealing to the *Bible*, uncomfortable. Nevertheless, the antebellum United States supported a variety of scientists and quasi-scientists who were happy to give lectures explaining to their audiences the physical justification for slavery.

Although emancipation of the slaves by 1865 rendered much of the controversy moot, the polygenist theory lingered on into the twentieth century. In the early years of the century, professional anthropologists attacked and crippled the theory among most social scientists. The reaction to Nazi atrocities during the 1940s weakened racist ideology outside the American South. Genetic research during the 1960s established beyond doubt that all humans are of the same species.[48] No responsible, informed adult these days would suggest that there is more than one human species.

Whether there are such things as human subspecies and whether some of those subspecies have different, and perhaps superior, qualities to others are still open questions in natural and social science.

(3) *Eugenics*. In the opening chapter of *The Origin of Species*, Darwin discussed the wide variations that humans had been able to breed into domestic plants and animals. He dwelt, particularly, on the many different races of domestic pigeon that had been created by human intervention in the process of reproduction. By starting with small natural variations in the wild rock pigeon and pairing males and females with complementary features—a variation in habits, voice, coloring, foot size or shape, beak size or shape, feather pattern, size, and numbers, even number of vertebrae—humans had created a profusion of types.[49] For the rest of the book, Darwin argued that what Man had done deliberately over a relatively short period of time with this one species, Nature had been doing blindly over eons with all species. Environmental forces, by "selecting" some variations and culling others, had caused various races to diverge so far that they became unable to interbreed, and had thus evolved into separate species.

Although he was too prudent to discuss the subject in the *Origin*, the most startling implication of Darwin's theory was immediately obvious to everyone: the history of *Homo sapiens* consisted of the emergence from previous species. Like all other creatures and plants, Man was an organism with an evolutionary history. But another, equally important implication did not take much longer to be realized: If humans had a biological past like that of other living things, then humans could similarly have a directed future. If selective breeding could create pigeons with superior speed or beauty, or a different type of coo, then targeted reproduction of humans could create either superior or inferior varieties of men and women.

By the time he published *The Descent of Man* in 1871, Darwin had lost his reluctance to explore the subject of human evolution. With caution but determination, he drew out the effects of selection, natural or otherwise, on the species. He argued that many different human qualities have been enhanced, in one group or another, by the different environmental situations in which they found themselves. As for the species as a whole, it is "highly probable that with mankind the intellectual faculties have been gradually perfected through natural selection."[50]

But if the human species had advanced toward perfection, it could also regress, and from the same causes. Whereas Nature tends to ruthlessly eliminate the inferior individuals of a species, "we build asylums for the imbecile, the maimed, and the sick; we institute poor-laws; and our medical men exert their utmost skill to save the life of every one to the last moment.... Thus the weak members of civilized societies propagate their kind."[51] The clear implication of his own theory was that cultural factors were substituting for Nature, and in fact selecting the wrong characteristics to increase in the population. The result must be that the species would deteriorate. But by the same logic, if humans could take control of their own breeding and discipline it along self-consciously Darwinist principles, they could reverse the deterioration and even create an evolutionary momentum toward a better *Homo sapiens*.

Darwin rejected the idea that any sort of government agency might be authorized to direct breeding programs to keep humanity fit, because such a policy would outrage morality. But others were not so squeamish about exploring the implications of the theory for the human future.

Darwin's cousin, Francis Galton, began advocating directed breeding during the 1870s, and coined the term "eugenics," from the Greek words for "well born," in 1883.[52] Whereas Spencer had been politically conservative, attempting to defend capitalism from the envy of the propertyless, Galton's motivations were pro-democracy. Because the "stupidity and wrong-headedness of many men and women [is] so great as to be scarcely credible," he opined, popular government was constantly at risk. But if "degenerates" could be prevented from having children, then the quality of the citizenry could be gradually raised, and the policies produced by self-government would improve.[53]

The idea had an intrinsic appeal to reformers. By the early twentieth century, eugenics was a popular cause among crusaders on the "progressive" side of the political spectrum in Europe and the United States. H. G. Wells, George Bernard Shaw, John Maynard Keynes, Oliver Wendell Holmes, Theodore Roosevelt, and Margaret Sanger were just a few of the luminaries of progressive politics who advocated controlled breeding at some time in their careers.[54] In the first decades of the century, compulsory sterilization laws for the feeble-minded were passed in thirty American states, Canada, the Scandinavian countries, and Germany.[55]

From a strictly scientific point of view, the advocates of eugenics were correct. If a faster pigeon or racehorse could be bred through the application of Darwinist principles to human reproduction, then a more intelligent—even, perhaps, a more moral—human might be created in the same way. But eugenicists were (rather surprisingly, in the case of Roosevelt) as politically naïve as they were socially optimistic. The classic mistake of progressives in the first three decades of the twentieth century was to believe that a group of technical experts in power would have no interests of their own but would be entirely honest and public spirited. The progressives were curiously confident that an elite of Platonic guardians could be appointed who would make reproductive decisions for society in a wholly disinterested manner. Although they placed no faith in businessmen, elected politicians, or judges, they kept coming up with classes of people who could be trusted to govern society in general, and presumably, reproduction in particular. Economist Thorstein Veblen, for example, recommended that political power be placed in the hands of industrial engineers.[56]

The eugenic delusion, that the right people could be found to decide who would be permitted to breed, came to grief, of course, with the Nazis. After 1945, the revulsion against the consequences of the German idea of a potentially master race discredited the project of eugenics so thoroughly that, more than sixty years later, it is still an epithet. As philosopher Philip Kitcher puts it, to even consider the "role of molecular genetics in designing future generations" is now difficult, because "the stigma associated with 'eugenics' is powerful enough to end the conversation right there."[57] Nevertheless, he predicts that recent advances in molecular biology, especially the Human Genome Project, will revive the discussion. "[A]s we envisage the many kinds of genetic tests that will become possible in the next decades, it is easy to fear that a benign policy of forestalling disease may become a program for enforcing social prejudice, a new eugenics."[58]

The ethical issues created by humanity's growing ability to replace "biological chance with technological choice," as philosopher Benjamin Gregg phrases it, are multiplying.[59] Gregg has concentrated on the question, "Might a human embryo be culturally understood as possessing a human right to be free of genetic manipulation?"[60] His answer is that political communities, or at least the elites of those communities, will make individual decisions about the issue. That is, whether an embryo has human rights will be socially determined. That is one position. But as modern science expands on the original insights from Darwin, Wallace, and Mendel, the controversies over what used to be called eugenics are going to become more frequent and intense.

Thus, although Hitler made "eugenics" an epithet, he did not eliminate the implications of the theory of natural selection for human reproduction.

(4) *Male and female.* For the first century or so after Darwin, evolutionary biologists largely studied the "morphology"—the outward form of the species in the history of life. Fossils were the evidence that organisms had evolved, and fossils were known by those parts of creatures that could be preserved in the geological record: shells, bones, teeth. Fossilized soft parts of bodies were rare. Not just rare but almost completely absent was direct evidence of behavior. Paleontologists could indirectly infer conclusions about behavior from morphology. A fossilized animal with webbed feet, for example, was assumed to have lived at least part of its life in the water. But because the behavior of dead creatures could not be directly observed, its study was a small part of the life of a paleontologist.

Meanwhile, the study of living creatures ("natural history") was a pastime that long predated Darwin, but even after the *Origin* was published, the activity was largely unsystematic and non-scientific. In the 1930s, such researchers as Tinbergen, Lorenz, and von Frisch began to apply scientific methodology to the study of animal behavior, thus creating the discipline of "ethology," but they did so without adopting a Darwinian perspective. In 1975, however, Edward O. Wilson, an entomologist by profession, combined the present-oriented perspective of ethology with the past-oriented perspective of paleontology, subsumed both under the theoretical approach of the Modern Synthesis of evolutionary biology, and wrote *Sociobiology*, thus presenting the world with the Darwinian study of behavior.

"In a Darwinian sense," wrote Wilson, "the organism does not live for itself. Its primary function is not even to reproduce other organisms; it reproduces genes, and it serves as their temporary carrier.... Sociobiology is defined as the systematic study of the biological basis of all social behavior."[61] Thus, the way to explain, say, the willingness of many animals, such as ants, to sacrifice themselves for the benefit of others of their kind ("altruism") is to ask how such a sacrifice benefits the genes that are shared between the sacrificer and the beneficiary of the sacrifice.[62]

Sociobiology instantly became a smash hit in the life sciences. And if Wilson had been content to illustrate the theory with references to ants, aardvarks, and iguanas, he would probably not have stirred a political tempest. But Wilson, unlike Darwin, had not been inhibited in pointing out the implied applications of his theory to *Homo sapiens*. "There is a need for a discipline of anthropological genetics," he wrote in the last chapter. "We are compelled to drive toward total knowledge, right down to the levels of the neuron and the gene. When we have progressed enough to explain ourselves in these mechanistic terms, and the social sciences come to full flower, the result might be hard to accept."[63]

Hard to accept, indeed. Wilson was clearly aware of the discomfort that a call for a science that would explain human behavior in terms of the interactions of genes and environment might create, but he underestimated the force of the reaction and did not anticipate its political origins, for the implications of an attempt to explain social life in terms of the interplay of nature and nurture are a direct, intense challenge to the most fundamental and cherished values of those on the left side of the political spectrum.

The primary value of those on the political left is *equality*.[64] Therefore, the primal social evil is a hierarchy of resources, status, and power that is imposed on the majority of the public by the elite that has the most of all those good things. As political scientist Harold Lasswell put it in the 1920s, politics, most of the time, is about who-gets-what-when-and-how.[65] In the view of people on the left, the inequalities in the distribution of who-gets-what are inflicted by the elite on the mass by various kinds of force and subterfuge. One of the chief forms of subterfuge is to make the existing inequalities appear to be sewn into the natural order, rather than artifacts of elite manipulation. As the French mathematician the Marquis de Condorcet expressed the idea during the eighteenth century, conservatives try to "make nature herself an accomplice in the crime of political inequality."[66] That, of course, is exactly what Spencer attempted to do with the theory of natural selection during the nineteenth century.

To the left, therefore, all efforts to "naturalize" inequalities are politically anathema and scientifically fraudulent. But leftists have an additional, different-but-related reason for detesting genetic explanations of human behavior. Leftists are reformers; they want to change society to be more just and equal. The more human behavior is the result of culture, the more society is amenable to improvement, because culture can be changed. But the more human behavior is natural, or, as leftists like to say it, "genetically determined," the less the hierarchies of society can be altered, and the more impossible appears their hope of changing things for the better. As Richard Lewontin expressed the idea in a conversation with me, "The world must be malleable; human nature cannot be fixed. A reformist person must believe that changing people is possible."[67] A reformist person therefore prefers that nurture accounts for all of the variation in human behavior, leaving nothing for nature to explain.

Wilson did not claim, in *Sociobiology* or subsequent publications, that the genes "determined" behavior and that therefore it would be futile to try to reform society. He contented himself with criticizing "the extreme orthodox view of environmentalism . . . that in effect there is no genetic variance in the transmission of culture" and suggesting that further research would uncover interactions between nature and nurture, that is, genetics and culture.[68] But to the leftists within the academy, *any* concession to nature was a retreat in the good fight for social reform, and a potential return to the bad old days of Spencerism.

The result was a political explosion in the life and social sciences. The gaudy saga of sociobiology over the next two decades after 1975 has been recounted and analyzed from various perspectives in a variety of venues, and I am not going to go over the entire ground again.[69] But the section of the sociobiological controversy that has lasted the longest, and can still provoke intense conflict nearly four decades after Wilson introduced the term, is the fight over the natural, or socially constructed, difference, or lack of difference, between men and women.

In the late 1970s, some psychologists took up Wilson's suggestion that they apply the theory and methods of sociobiology to human behavior, and the subfield of "evolutionary psychology" was born. Scholars in this subfield start from the basic sociobiological assumption that, in John Alcock's words, "the brain is essentially a reproductive organ, like every other evolved internal mechanism of living beings."[70] Evolutionary psychologists then add an additional assumption: that the human mind became adapted to the conditions of the Pleistocene epoch of roughly 200,000 to 10,000 years ago and has not had time to catch up to the conditions of modern, urban, technological society.[71] Thus, the "Stone-Age mind"[72] we all use to get through life is still employing the cognitive and emotional strategies that were appropriate when we lived in small—thirty would be unusually large—hunter-gather groups in which we knew everyone personally and were probably kin to the other members of the group.

Evolutionary psychologists rarely address political questions directly, although a few political scientists have applied their methodology to traditional questions of power, legitimacy, and mass behavior.[73] Nevertheless, the difference between male and female in regard to emotions and cognition is one of their most popular topics, and the implications of their conclusions have major political consequences.

Although evolutionary psychologists usually assert that they are not attempting to explain *individual* behavior,[74] they tend to write in broad generalizations that submerge the potential uniqueness of human personality. Moreover, the generalizations they promote are almost always restatements of the crudest kinds of sexual stereotypes. Thus, in the portrait of the sexes painted by sociobiologists, men are (or want to be) promiscuous in their choice of sexual partners, while women are choosy. Men commit crimes, while women are law-abiding. Men are competitive, while women are cooperative. Men are physically and symbolically aggressive, while women prefer to use peaceful negotiation to resolve conflict. Men control political and economic society, while women control the home environment. Men like to play games, while women like to go shopping. Women enjoy taking care of children, while men do it as a duty.[75] As psychologist David Barash has incautiously summarized, "If male-female differences are sexist, we should put the blame where it really belongs, on the greatest sexist of all: 'Mother' Nature!"[76]

Leftists in and outside of the universities have reacted with fury to the evolutionary psychologists' generalizations about the sexes. Their first criticism is to point out that evolutionary psychologists frequently draw their conclusions from—that is, claim that their theories are supported by—seriously inadequate evidence. A variety of publications have appeared explicating the rather sloppy use of data to be encountered in evolutionary psychology's publications.[77]

But despite the importance of the quality of the treatment of evidence in all social science, the second leftist criticism of the application of sociobiology to the problem of male and female is more fundamental. This is the charge that any research based on the premise that men and women are significantly different is actually a masked defense of all the unjust hierarchies of modern society, especially patriarchy. In Bonnie Spanier's summary, E.O. Wilson's "arguments rationalized the cultural dominance of white, heterosexual males" and helped to "underpin justifications for sexism, racism, classism, and heterosexism."[78] As biologist Ruth Bleier put it, sociobiology is an attempt to "implicitly defend the status quo" by taking differences that are socially constructed and making them appear to be part of the natural order.[79]

To the left, then, it is not so much the habitually imperfect conduct of science that condemns sociobiology/evolutionary psychology as an enterprise. After all, inadequate scientific technique can always be improved. What can never be improved is the implicit political meaning of an approach that must interpret invidious distinctions between male and female as products of an unalterable evolutionary history rather than a reformable cultural pattern.

The sociobiologists and evolutionary psychologists who are thus being accused of committing a scientific crime by implication deny vociferously that they are political right-wingers. More profoundly, they deny that what they are doing is political at all. Like the members of the Royal Society before them, they argue quite vigorously that they are only fair-minded scientists attempting to discover the truth about nature. They believe that their critics may be political but that they are not. "I had no interest in ideology. My purpose was to celebrate diversity and to demonstrate the intellectual power of evolutionary biology," wrote Wilson in his memoirs.[80] In her account of the controversy, Ullica Segerstrale, who is both a sociologist and a biochemist, accused Wilson's critics of committing a "moral reading" of science, of having a "coupled agenda," and of pursuing "politics by scientific means."[81] Many other voices have been raised, defending the practice of evolutionary biology as simply an honest effort to learn about human nature, with no political agenda other than the pursuit of truth.[82]

The battle continues to rage in the second decade of the twenty-first century, and undoubtedly will continue to do so when all of the recent combatants have left the arena. Should science in general and biology in particular be viewed as the disinterested search for truth, which might justify letting those who engage in it go about their business without interference? Or should they be viewed as ineradicably infected with political ideologies, which might justify subjecting their activities to some sort of social supervision?

Since the argument has been ongoing for centuries, it is unlikely that the conflict will be settled soon. I write from the stance of one who embodies the conflict within his own thoughts. I believe that objective reality exists, that science is the best method yet devised for discovering what it is, and that the optimum strategy for persuading scientists to discover the nature of that reality is to give them money and then leave them alone. But I also believe that most science—and this goes double for biology, and triple for evolutionary biology—has political implications. I justify my own inability to resolve the conflict by consoling myself that no one else has been able to resolve it either.

This uncomfortable situation of perpetual tension within science, and within those who think about science, is simply one aspect of the contradictory essence of the politics of evolution. Not everybody who takes a position on the issues surrounding evolution, and especially a position on the particular flashpoint of government policy, the public school biology curriculum, is intellectually conflicted. Indeed, some are clear and certain to the point of zealotry. But within society, the zealotries line up against each other, so that the political situation as a whole is one of annoyed deadlock.

Conflicts occur within the arena of practical politics, and I will get to those. But first, I want to explore some types of conflict within the discourse of evolution. Science, much more than politics, is heavily invested in the dispassionate evaluation of evidence. But, like politics, it is also dependent on the uses of rhetoric. The employment of a particular form of rhetoric—metaphor—has been of great importance to both the inner biological dialogue and the outer biological controversy.

## Notes

1 Stone, Deborah, *Policy Paradox: The Art of Political Decision Making.* 3rd ed. (New York: W.W. Norton, 2012), 2.
2 Dennett, Daniel, "Why Getting It Right Matters," in Paul Kurtz, ed., *Science and Religion: Are They Compatible?* (Amherst, N.Y.: Prometheus Books, 2003), 149–159.
3 I discuss the artistic personality in more detail in *The Politics of Glamour: Ideology and Democracy in the Screen Actors Guild* (Madison: University of Wisconsin Press, 1988) and *Risky Business: The Political Economy of Hollywood* (Boulder, Colo.: Westview Press, 1993).
4 This theme is explored in greater detail in Ferris, Timothy, *The Science of Liberty: Democracy, Reason, and the Laws of Nature* (New York: HarperCollins, 2010).
5 Bacon, Francis, *Magna Instauratio* (originally published 1620), in Richard Foster Jones, ed., *Francis Bacon: Essays, Advancement of Learning, New Atlantis, and Other Pieces* (New York: Odyssey Press, 1937), 308, 310.
6 Bacon, Francis, *The Advancement of Learning* (originally published 1605) in Jones, ibid., 214.
7 Gleick, James, "At the Beginning: More Things in Heaven and Earth," in Bill Bryson, ed., *Seeing Further: The Story of Science, Discovery, and the Genius of the Royal Society* (New York: HarperCollins, 2010), 27–28; Barnes, Barry, David Bloor, and John Henry, *Scientific Knowledge: A Sociological Analysis* (Chicago: University of Chicago Press, 1996), 145.
8 Kaplan, Abraham, *The Conduct of Inquiry: Methodology for Behavioral Science* (Scranton, Pa.: Chandler Publishing Company, 1964), 381.
9 Miller, Kenneth R., *Only a Theory: Evolution and the Battle for America's Soul* (New York: Penguin, 2008), 201.
10 Sesardic, Neven, "Nature, Nurture, and Politics," *Biology and Philosophy*, 25 (2010), 433–436.
11 Maienschein, Jane, "What Is an 'Embryo' and How Do We Know?" in David L. Hull and Michael Ruse, eds., *The Cambridge Companion to the Philosophy of Biology* (Cambridge: Cambridge University Press, 2007), 341; Brown, Mark B. *Science in Democracy: Expertise, Institutions, and Representation* (Cambridge, Mass.: MIT Press, 2009), 198.
12 Ruse, Michael, "Evolutionary Ethics: Healthy Prospect or Lost Infirmity?" in Matthew Mohan and Bernard Linsky, eds., *Philosophy and Biology* (Chicago: University of Chicago Press, 1988), 58.
13 Weber, Max, "Value-judgments in Social Science," in Richard Boyd, Philip Gasper, and J.D. Trout, eds., *The Philosophy of Science* (Cambridge, Mass.: MIT Press, 1991), 719.
14 Rosenberg, Alex and Daniel W. McShea, *Philosophy of Biology: A Contemporary Introduction* (New York: Routledge, 2008), 221.
15 Greenberg, Daniel S., *Science, Money, and Politics: Political Triumph and Ethical Erosion* (Chicago: University of Chicago Press, 2001), 44 and passim; Spanier, Bonnie B.,

*Im/partial Science: Gender Ideology in Molecular Biology* (Bloomington: Indiana University Press, 1995), 128–133.
16. Shattuck, Roger, *Forbidden Knowledge: From Prometheus to Pornography* (New York: St. Martin's Press, 1996), 174–177.
17. Levins, Richard and Richard Lewontin, *The Dialectical Biologist* (Cambridge, Mass.: Harvard University Press, 1985), 5.
18. Cohen, H. Floris, *The Scientific Revolution: A Historiographical Inquiry* (Chicago: University of Chicago Press, 1994), 167.
19. Ibid.
20. For example: Pearcey, Nancy R., "Darwin Meets the Berenstain Bears: Evolution as a Total Worldview," in William A. Dembski, ed., *Uncommon Dissent: Intellectuals Who Find Darwinism Unconvincing* (Wilmington, Del.: ISI Books, 2004), 60; Coulter, Ann, *Godless: The Church of Liberalism* (New York: Three Rivers Press, 2007), 268–269; Yahya, Harun, *Fascism: The Bloody Ideology of Darwinism* (Istanbul: Arastirma Publishing, 2002), passim.
21. Moy, Timothy, "The Galileo Affair," in Kurtz, *Science and Religion*, op. cit., 139–144.
22. Feyerabend, Paul, *Against Method* (London: Verso, 2010), 127–134.
23. Hobbes, Thomas, *Leviathan*.
24. Shapin, Steven and Simon Schaffer, *Leviathan and the Air-Pump: Hobbes, Boyle, and the Experimental Life* (Princeton, N.J.: Princeton University Press, 1985), 325.
25. Ibid., 107, 109.
26. Weinberg, Steven, *Facing Up: Science and Its Cultural Adversaries* (Cambridge, Mass.: Harvard University Press, 2001), 33.
27. Angier, Natalie, *The Canon: A Whirligig Tour of the Beautiful Basics of Science* (Boston: Houghton Mifflin, 2007), 242–243.
28. Weinberg, Steven, *The First Three Minutes: A Modern View of the Creation of the Universe* (New York: Bantam Books, 1977), 2.
29. Stenger, Victor, *Has Science Found God? The Latest Results in the Search for Purpose in the Universe* (Amherst, N.Y.: Prometheus Books, 2003), 37.
30. Strobel, Lee, *The Case for a Creator: A Journalist Investigates Scientific Evidence that Points toward God* (Grand Rapids, Mich.: Zondervan, 2004), 77.
31. Dembski, William, *Intelligent Design: The Bridge between Science and Theology* (Downers Grove, Ill.: InterVarsity Press, 1999), 203.
32. Brown, Walt, *In the Beginning: Compelling Evidence for Creation and the Flood*, 8th ed. (Phoenix, Ariz.: Center for Scientific Creation, 2008), 31.
33. Faulkner, Danny, *Universe by Design: An Explanation of Cosmology and Creation* (Green Forest, Ark.: Master Books, 2004), 5, 96.
34. Spencer, Herbert, *Social Statics: The Conditions Essential to Human Happiness Specified and the First of Them Developed* (n.p.: Forgotten Books, 2012), 378; first published 1850.
35. I am, of course, using teleological language here, and not explaining the overproduction of offspring the way a biologist would. But for the purposes of introducing and summarizing Darwin's theory, such usage is acceptable. For a discussion of this general topic, see Ruse, Michael, *Darwin and Design: Does Evolution Have a Purpose?* (Cambridge, Mass.: Harvard University Press, 2003), 8, 206, and passim.
36. Darwin, Charles, *On the Origin of Species by Means of Natural Selection* (New York: Barnes and Noble, 2004).
37. Bowler, Peter J., *Evolution: The History of an Idea*, 3rd ed. (Berkeley: University of California Press, 2003), 260–264.

38 Mayr, Ernst and William B. Provine, eds., *The Evolutionary Synthesis: Perspectives on the Unification of Biology* (Cambridge, Mass.: Harvard University Press, 1998); Smocovitis, Vassiliki Betty, *Unifying Biology: The Evolutionary Synthesis and Evolutionary Biology* (Princeton, N.J.: Princeton University Press, 1996).
39 Spencer, Herbert, *Social Statics: The Conditions Essential to Human Happiness Specified and the First of Them Developed* (n.p.: Forgotten Books, 2012); first published 1850.
40 Spencer, Herbert, *The Man Versus the State* (Indianapolis: Liberty Fund, 1982); originally published 1884.
41 Hofstadter, Richard, *Social Darwinism in American Thought* (Boston: Beacon Press, 1944, 1992); Prindle, David, *The Paradox of Democratic Capitalism: Politics and Economics in American Thought* (Baltimore: Johns Hopkins University Press, 2006), 115–116, 124, 138, 152–154, 159; Werth, Barry, *Banquet at Delmonico's: Great Minds, the Gilded Age, and the Triumph of Evolution in America* (New York: Random House, 2009).
42 On Sumner's influence, see my *Paradox of Democratic Capitalism*, 110–113, 116, 131.
43 Prindle, *Paradox of Democratic Capitalism*, op. cit.
44 I discuss this point in greater detail in Prindle, David, *Stephen Jay Gould and the Politics of Evolution* (Amherst, N.Y.: Prometheus Books, 2009), 66–71, 207.
45 Himmelfarb, Gertrude, *Darwin and the Darwinian Revolution* (Chicago: Ivan R. Dee, 1996), 431.
46 The information in this section is based on material in the following: Desmond, Adrian and James Moore, *Darwin: The Life of a Tormented Evolutionist* (New York: W.W. Norton, 1991), 521; Bowler, *Evolution*, op. cit., 293–296; Degler, Carl, *In Search of Human Nature: The Decline and Revival of Darwinism in American Social Thought* (Oxford: Oxford University Press, 1991), 101, 204; Gould, Stephen Jay, *The Panda's Thumb: More Reflections in Natural History* (New York: W.W. Norton, 1980), 169–176; Gould, Stephen Jay, *The Mismeasure of Man*, 2nd ed. (New York: W.W. Norton, 1996), 70–101.
47 Gould, *Mismeasure*, 84–87.
48 Ruvolo, Maryellen and Mark Seielstad, "The Apportionment of Human Diversity 25 Years Later," in Rama S. Singh, Costas B. Krimbas, Diane B. Paul, and John Beatty, eds., *Thinking about Evolution: Historical, Philosophical, and Political Perspectives* (Cambridge: Cambridge University Press, 2001), 141–151.
49 Darwin, Charles, *On the Origin of Species*, 1st ed. (New York: Barnes and Noble Classics, 1859, 2004), 17–45.
50 Darwin, Charles, *The Descent of Man* (New York: W.W. Norton, 1871, 2006), 868.
51 Ibid., 873.
52 Degler, Carl N., *In Search of Human Nature: The Decline and Revival of Darwinism in American Social Thought* (New York: Oxford University Press, 1991), 41.
53 Galton quoted in Ferris, *Science of Liberty*, op. cit., 30.
54 Pinker, Steven, *The Blank Slate: The Modern Denial of Human Nature* (New York: Viking, 2002), 153.
55 Ibid., 16.
56 I discuss this point at greater length in Prindle, *The Paradox of Democratic Capitalism*, op. cit., 162–168.
57 Kitcher, Philip, "Utopian Genetics and Social Inequality," in *In Mendel's Mirror: Philosophical Reflections on Biology* (Oxford: Oxford University Press, 2003), 261.
58 Ibid., 262.
59 Gregg, Benjamin, *Human Rights as Social Construction* (Cambridge: Cambridge University Press, 2012), 209.

60  Ibid., 190.
61  Wilson, Edward O., *Sociobiology: The New Synthesis*, 2nd ed. (Cambridge, Mass.: Harvard University Press, 2000), 3–4.
62  Ibid., 421–428.
63  Ibid., 550, 575.
64  I explore this point in detail in *The Paradox of Democratic Capitalism*, op. cit., passim, and in *Stephen Jay Gould and the Politics of Evolution*, op. cit., 68–69, 84–85, 150.
65  Lasswell, Harold, *Politics: Who Gets What, When, How* (New York: Meridian Books, 1936, 1958).
66  Condorcet quoted in Gould, *Mismeasure*, op. cit., 53.
67  DFP phone conversation with Richard Lewontin, October 31, 2006.
68  Wilson, *Sociobiology*, op. cit., 550.
69  Segerstrale, Ullica, *Defenders of the Truth: The Sociobiology Debate* (Oxford: Oxford University Press, 2000), passim; Wilson, Edward O., *Naturalist* (New York: Warner Books, 1994), 307–353; Prindle, *Stephen Jay Gould and the Politics of Evolution*, op. cit., 131–141.
70  Alcock, John, *The Triumph of Sociobiology* (Oxford: Oxford University Press, 2001), 150.
71  Cosmides, Leda, John Tooby, and Jerome Barkow, "Introduction: Evolutionary Psychology and Conceptual Integration," in Barkow, Cosmides, and Tooby, eds., *The Adapted Mind: Evolutionary Psychology and the Generation of Culture* (New York: Oxford University Press, 1992), 5.
72  Buller, David J., *Adapting Minds: Evolutionary Psychology and the Persistent Quest for Human Nature* (Cambridge, Mass.: MIT Press, 2005), 59–60.
73  Somit, Albert and Steven Peterson, *Darwinism, Dominance, and Democracy: The Biological Basis of Authoritarianism* (Westport, Conn.: Praeger, 1997); Masters, Roger D., *The Nature of Politics* (New Haven, Conn.: Yale University Press, 1989); Saad, Gad, "Evolution and Political Marketing," in Albert Somit and Steven Peterson, eds., *Human Nature and Public Policy: An Evolutionary Approach* (New York: Macmillan, 2003), 121.
74  Pinker, *The Blank Slate*, op. cit., 340; Cosmides, Leda and John Tooby, "The Psychological Foundations of Culture," in Barkow et al., eds., *The Adapted Mind*, op. cit., 66.
75  Buss, David M., *Evolutionary Psychology: The New Science of the Mind* (Boston: Pearson, 2004), 103–186; Buss, David M., *The Evolution of Desire: Strategies of Human Mating* (New York: Harper/Collins, 1994), passim; Wilson, Edward O., *On Human Nature* (Cambridge, Mass.: Harvard University Press, 2004), 121–148; Pinker, *The Blank Slate*, op. cit., 316–317, 340–346, 353–356.
76  Barash, David, *The Whisperings Within: Evolution and the Origin of Human Nature* (New York: Penguin, 1979), 90.
77  Buller, *Adapting Minds*, op. cit.; Prindle, *Stephen Jay Gould and the Politics of Evolution*, op. cit., 135–138.
78  Spanier, *Im/partial Science*, op. cit., 33.
79  Bleier, Ruth, *Science and Gender: A Critique of Biology and Its Theories on Women* (Oxford: Pergamon Press, 1984), vii.
80  Wilson, Edward O., *Naturalist* (New York: Warner Books, 1994), 336.
81  Segerstrale, *Defenders of the Truth*, op. cit., 2, 41–42, 77, 120, 206.
82  Pinker, *The Blank Slate*, op. cit., viii; Alcock, *The Triumph of Sociobiology*, op. cit., 190, 192.

# 2
# EVOLUTION AND METAPHOR

Evolution is "the greatest show on Earth,"[1] and the scientists who study it are the "architects of eternity."[2] Furthermore, it is "the canary in the mineshaft, an indicator whose presence signals the health or sickness of the entire scientific enterprise."[3] Meanwhile, don't be fooled; the claim of the partisans of "Intelligent Design" (ID) is a scam. ID is just "creationism's Trojan horse."[4] In fact, it is merely "creationism in a cheap tuxedo."[5]

But wait. Actually, Darwinism is a disease that has "infected the whole culture," producing everything from "dictators to today's ... sexual profligates, and animal rights nuts."[6] Or, it is like a monster with "evil tentacles" that "have crept into all corners of modern thought."[7] Presently, it has a "stranglehold" on scientific education. But, thankfully, that stranglehold will soon be broken, because it is a "crumbling edifice"; in fact, the only reason it still survives is because of "smoke and mirrors,"[8] or perhaps because of the actions of the "Darwinian thought police."[9]

As the above two paragraphs illustrate, discussing the theory of natural selection without employing metaphors is difficult. Perhaps it is impossible.

In this chapter I will examine Darwinism and metaphor from three points of view. First, I will examine the uses of metaphor in human thought in general and scientific thought in particular. Second, I will analyze the careers of two metaphors that Darwin employed in *The Origin of Species*. Third, I will discuss the use of metaphor within the polemics that surround the issue of Darwinism in the larger society.

## A Man's Reach Should Exceed His Grasp—or What's a Metaphor?

The most basic kind of cognitive tool is the "concept." A concept is an entity of understanding that "carves at the joints" of reality,[10] "an abstraction representing an object, a property of an object, or a certain phenomenon,"[11] a "unit of thinking."[12] It is, then, a representation in words of something that a thinker believes is present in empirical reality and that therefore functions as a mental tool to permit the researcher to try to discover a truth about the world.

An "analogy" is a specific kind of concept. It is a comparison between two things, based on the assumption that "if they agree with one another in some respects they will probably agree in others,"[13] permitting some inferences about an unknown entity based on knowledge about a known entity. (For example, the flow of electrons through a wire is often analogized to the movement of water in a confined space, and studied using the same cognitive tools.)[14]

A metaphor is a specific type of analogy, a figure of speech that makes a comparison. As with many another important term, "metaphor" is not always defined the same way by everyone who writes about it nor always used the same way in expression. The boundary between analogies and metaphors is so hard to pin down that some scholars use the two concepts interchangeably.[15] For our purposes, a metaphor is a concept expressed in language that symbolically compares one thing to another or explains one thing in terms of another. "I think that in this war we can see the light at the end of the tunnel." "His thesis was buried under an avalanche of criticism." "The job of the Fed is to take away the punchbowl just as the party is getting good." And so on. Metaphors fit into the classical discipline called *rhetoric*, the practice of producing "messages designed to influence human thoughts and actions."[16] A rhetorician—or, in more familiar form, a poet, a spinner of yarns, an orator, a persuader, a demagogue, even a painter, cartoonist, dramatist, or choreographer—uses a metaphor to guide a listener's, reader's, or viewer's mind toward a particular cognitive interpretation and emotional state.

As the metaphors quoted in the first two paragraphs of this chapter illustrate, these figures of speech are well suited to polemical purposes. They fit into what might be termed a "rhetoric of evaluation." The quoted metaphors do not tell us anything factual about the world. Instead, they try to rearrange our thoughts and emotions into a configuration in which we either value (first paragraph) or reject (second paragraph) some concept, situation, or person. Scholars of the psychology of persuasion acknowledge that metaphor often plays a major role in such rhetoric, although they have not made much progress in discovering rules about the circumstances under which a metaphor is successfully apt and convincing.[17]

But metaphors can also serve more serious and educational ends. In science, for example, they might be said to fit into a "rhetoric of understanding." That is, a metaphor can be used to help people come to a more or less correct interpretation of the way the world works.

Early in the Scientific Revolution, when trying to discover and master the rules of clear thinking, intellectuals grew suspicious of metaphor, believing that it distracted the mind from rigorous reasoning into realms of poetic fluff. Newton referred to metaphor as "a kind of ingenious nonsense," and Bacon described it "rather as a pleasure or play of wit than a science."[18] Locke, convinced that much human conflict was caused by philosophers who "perplex and confound the signification of words, and thereby render the language less useful," warned in *An Essay Concerning Human Understanding* against the use of "obscure, doubtful, and undefined words," although he did not use the specific term "metaphor."[19]

Well into the twentieth century, the ideal of antiseptic analysis reported without rhetorical tropes such as metaphor dominated the self-concept of scientists and philosophers of science. As Brooks and Warren articulated the "windowpane theory" of scientific discourse in 1938:

> The primary advantage of the scientific statement is that of absolute precision.... Such precision ... can be gained only by using terms in special and previously defined senses. The scientist carefully cuts away from his scientific terms all associations, emotional colorings, and implications of attitude and judgment.[20]

Or as Davida Charney described the model of scientific prose in 1993:

> Since at least Bacon's time, scientists have taken as their ideal a form of scientific discourse that is straightforward, objective, and dispassionate, discourse that confines itself to describing independently confirmable observations and drawing dispassionately logical conclusions from them. As such, generations of scientists have conceived of their discourse as standing outside the realm of rhetoric, the classical art of persuasion.[21]

But the windowpane theory of scientific expression was never universal in application. Beginning in the 1970s, some scholars had begun arguing that metaphor was essential to thought in general and scientific communication in particular. They decided that human minds had great difficulty learning anything truly original, and that they best assimilated new understandings by comparing novel and mysterious knowledge to old and understood knowledge. A metaphor, argued George Lakoff and Mark Johnson in a groundbreaking 1980 book, "unites reason and imagination," which literally permit humans to think.[22] By the 1990s, it had become accepted that ordinary human language, or even thinking, would be impossible without metaphorical expressions.[23] In specific regard to the scientific project, modern scholars generally argue that "phenomena become objects of scientific discourse by virtue of the metaphor that makes them accessible to cognition"[24] and that "the very nature of science ... is such that scientists need the metaphor as a bridge between old and new theories."[25]

As is usual when they have studied a concept long enough, scholars came to see distinctions here, finding more than one metaphor type, with different types relating differently to the scientific project. Most relevantly, scholars of scientific metaphors distinguish between "pedagogical" metaphors, which help to *explain* a theory to outsiders or provide one possible *interpretation* of phenomena, and "theory constitutive" metaphors, which function as "indispensable parts of a scientific theory" because they "cannot be reformulated in literal terms."[26] An example of the first would be the Bohr atom in physics, pedagogically vivid as a little solar system but not telling scientists much about the actual relationships between atomic particles. An example of the second would be cognitive psychology's use of the term "feedback" to characterize human "information processing," "for which no adequate literal paraphrase is known."[27]

Arguments over the relationship between the different kinds of metaphors, and the various puzzles that philosophers of science and scholars of rhetoric have considered over the last five decades, have generated a large, convoluted, and often opaque literature. The major theme running through this literature is whether the language that scientists use to think about and describe their work tells us something true about the world or colors their thoughts and discussions such that the language itself impedes the scientific enterprise. As Thomas Kuhn summarized the problem in 1979, "Does it ... make better sense to speak of accommodating language to the world [rather] than of accommodating the world to language?"[28]

I will not be reviewing the many arguments that scholars have made while considering this theme.[29] I will simply admit to a rather old-fashioned leaning toward realism and empiricism, meaning that I believe that what distinguishes science from other human activities is its anchor in the physical world—both in and beyond laboratories. In my view, the most sensible statement by a philosopher of science in this area was Hilary Putnam's 1975 observation that the "positive argument for realism is that it is the only philosophy that doesn't make the success of science a miracle."[30]

Consequently, I see the need for another category of metaphor, one that embodies the transition between the mind's intuition, which is more likely to be mistaken, and the empirical testing which creates a knowledge that is less likely to be mistaken. This new category includes the necessity of *operationalization*, the definition of concepts in terms of observable, measurable indicators.[31] (Philosophers of science will want me to add here that empirical measurements must still be interpreted, and all interpretation involves language.) A metaphor that is not operationalized can still be a powerful aid to *thinking*, but to move it into the realm of scientific usefulness, it must be defined in a way that allows it to become measurable. Thus, metaphors may move back and forth between the realm of aids to thinking and the realm of scientific discourse, depending on their state of operationalization.

Two contrasting examples from the history of political science will help to make these distinctions clear. On the one hand, an example of the "pedagogical" type of metaphor would be Morton Grodzins' suggestion more than four decades ago that we should interpret federalism in the United States as being like a marble cake rather than a layer cake.[32] Virtually all discussions in textbooks adopt Grodzins' imagery; it has become part of the vocabulary of the discipline. However, no one has operationalized the two types of federalism, so the terms remain merely helpful non-scientific aids to understanding. On the other hand, McCubbins and Schwartz's 1984 admonition that congressional oversight of administrative agencies should be conceptualized as consisting of "fire alarms" rather than "police patrols" was supported by a good deal of empirical evidence.[33] For all their evocative vividness, therefore, and for all the ways they originate in imagistic discourse, the two concepts have an empirical grounding and have therefore journeyed from "constitutive" metaphors into scientific concepts.

Because people, including scientists, use metaphors to express their meaning, and because people argue about meaning, they often argue about metaphors. For example, in the real world of American politics, in late 2012 and early 2013, public intellectuals wrangled over the "right" metaphor for a policy conflict. When fighting over whether to approve payments of public debts incurred by previous Congresses during the summer of 2011, Republicans in the House of Representatives threatened to vote against the debt payment bill, perhaps wrecking the entire economy, unless Democrats in the Senate and the White House agreed to sweeping budget cuts for domestic programs. For their part, Democrats argued that payment of the debts for past spending, and decisions about present and future spending, should not be conflated but should be considered separately. The two sides compromised by agreeing to set themselves a deadline of the first week of January 2013, after which, if an overall spending policy had not been achieved, income taxes on everyone would automatically be raised and draconian across-the-board spending cuts would be imposed on all government programs.

As the deadline approached in late 2012, commentators and politicians wrestled with a metaphoric term to describe what would happen if an agreement was not reached. At first the verbal image "fiscal cliff," with its connotations of sudden fatal catastrophe, was the metaphor applied by everyone in politics and the media. But as weeks passed, and as Republicans and Democrats threatened each other with being held responsible for economic destruction, some observers began to argue that "cliff" was too drastic a metaphor and should be replaced with something still clearly ominous but not so instantly terminal.

Various writers suggested that a fiscal "slope" or "hill" would be more appropriate, because the damage done to the economy by budget cuts would be gradual rather than instantaneous.[34] Others, arguing that the "cliff" metaphor had simply become tiresome, suggested that new phrases such as "red ink stink," "president's precipice," or "dollar ditch" would be more agreeable.[35] After a new compromise was reached deferring hard budget decisions for a few more months the night before the 2011 agreement was to be implemented, editorial cartoonist Jack Ohman asked, in a six-panel cartoon, "We've avoided the Fiscal Cliff . . . Now What? Fiscal fission (over a drawing of an atomic explosion)? Fiscal tsunami (over a drawing of a monstrous ocean wave)? Fiscal earthquake (over a drawing of buildings crumbling)? Fiscal tornado (over a drawing of a funnel cloud heading toward a farm)?" In the sixth panel, Ohman pictured a lament coming out of Congress: "We can't even agree on a metaphor."

The reason that people care about which metaphor is most appropriate is that a metaphor applied to a problem implies a solution to the problem. As Deborah Stone observes, "Embedded in every policy metaphor is an assumption that if *a* is like *b*, then the way to solve *a* is to do what you would do to solve *b*."[36] Thus figures of speech become policy prescriptions.

Not only pundits but also scholars clash over the appropriateness of specific metaphors. For example, for much of the century or so after Darwin, controversy persisted over whether an animal's embryo was more accurately depicted as a "blueprint" or a "recipe." In the blueprint case, a paper telling a builder what to build is expected to correspond closely, even precisely, to whatever building the builder finally does build. In the recipe case, a set of instructions, if followed, is expected to result in a cake, but the cake will not look like the recipe. When in 1953 the structure of DNA was first described, emphasis began to shift away from embryology and toward genetics, and the genetic-recipe metaphor came to dominate biological thinking.[37]

In the 1980s, however, an additional discussion arose among social theorists about the appropriateness of these same two metaphors, a discussion explicitly directed at the political implications of each image. Some feminist critics argued that the genetic-blueprint metaphor was masculine because it encouraged control over a passive subject rather than allowing for creative variation, as a recipe does. More prosaically, in real life, men had more experience with blueprints and women with recipes, so the blueprint metaphor would "gender" public understanding disadvantageously, while the recipe metaphor would prove progressive.[38]

This is one of the few times when a dispute about metaphors was subjected to empirical tests. Celeste Condit and her collaborators measured public understanding of the two visions of genes through analysis of opinion surveys, press reports, and answers to questions in focus groups. Their conclusion was that "there is little evidence that either metaphor is distinctively associated with any of the concepts that the critics identified." Non-scientists had many different interpretations of the two metaphors, and those interpretations varied wildly according to context. The authors were unable to come to any broad conclusions about how the public perceived the meaning of the metaphors or applied them to biology.[39]

Metaphors, then, are everywhere in science and politics, in discourse public and private, but their meaning and social impact are always subject to dispute. Because the theory of evolution has been a major component of the intellectual world from the nineteenth into the twenty-first centuries, any discussion of its role must include a discussion of the metaphors it has adopted and spawned and the way it has been understood metaphorically by intellectuals and the wider public.

## The Foundation

Metaphors function to help us *understand*, but they are also one of the language tools that humans use to *persuade*. In the case of science, it can be difficult to distinguish the two functions of metaphors because scientists are always trying to persuade other scientists, and sometimes the public, to understand the world their way. Science differs from common politics, however, in that persuasion depends, very importantly, on whether the conclusions being urged are supported by empirical reality. But in the group analysis of evidence that comprises much of the scientific enterprise, metaphors can be crucial in persuading other scientists to take an author's thesis seriously, or to look at evidence from a particular point of view. Conversely, many scientific theories—some interpreting new evidence, others reinterpreting old evidence—have been slow to catch on because the metaphors their authors have offered were not cogent, not clear, or not consistent. Both Charles Darwin and many of his critics were and remain among such authors.

The "windowpane" ideal of scientific exposition lasted longer than might have been expected, given that one of the two or three most important scientific books ever written violated it in spectacular fashion. Darwin's *On the Origin of Species by Natural Selection*, published in 1859, was full of metaphorical aids to understanding. Moreover, its central argument—so central that the author put it into its title—was consciously presented as a metaphor.

In his first chapter, Darwin described the way human breeders had produced a great variety of, or many "races" of, pigeons. These "races" had come to differ in structure and behavior through "the continuous selection of slight variations,"[40] so that "man adds them up in certain directions useful to him. In this sense he may be said to make for himself useful breeds."[41] In an analogous manner, he argued, nature had from among each species retained for breeding those individuals better adapted to their respective environments. As a result, "any variation, however slight and from whatever cause proceeding, if it be in any degree profitable to an individual of any species . . . will tend to the preservation of that individual, and will generally be inherited by its offspring."[42] Because "natural selection is daily and hourly scrutinizing, throughout the world, every variation, even the slightest, rejecting that which is bad, preserving and adding up all that is good" over "the long lapse of ages," the variations had turned into different species, and then even more species, this process resulting in the profusion of life seen in Darwin's day and, minus modern extinctions, our own.[43]

Darwin thus presented the central argument of his book as a metaphor—what humans have been doing consciously as pigeon fanciers for a few thousand years nature had been doing unconsciously for many million years. But reading him in his entirety, Darwin clearly intended the actions of the environment to be "un-selection-like." "To select" was a transitive verb; it implied choice and agency. Darwin's very purpose, however, was to refute the idea, held by virtually every naturalist who preceded him, that species had come into existence through the conscious agency of a human-like mind, God's mind. The success of some species and the extinction of others, he wanted us to understand, was unconscious and unintended; it was the residuum of uncountable interactions between organisms and blind circumstance. The passive sorting of a sieve would therefore have been a metaphor that more accurately conveyed his meaning. In fact, during the nineteenth century, biologists would frequently substitute the metaphor of a sieve for the metaphor of selection in their discussions of Darwin's theory.[44] But Darwin himself stuck with the original metaphor, causing a great deal of confusion.

Some scholars—both philosophers and natural scientists—argue that the comparison between artificial and natural selection is just plain wrong, that natural selection cannot account for the generation of new species.[45] They constitute a small minority among people who think seriously about evolutionary biology, but, as with any heresy, their ideas may one day become orthodoxy. One thing the critics of the selection metaphor have not devised, however, is a counter-metaphor. Thus, Fodor and Piattelli-Palmarini, after 152 pages criticizing Darwin's naming of evolution's mechanism as "selection," admit that "we don't know what the mechanism is."[46] They do not have a theory, and so they cannot have a metaphor to explain their theory.

It is often said that in science, as in politics, "you can't beat something with nothing." Unsatisfactory theories are not defeated by their own inadequacies but by better theories. Similarly reliable as a generalization is this: A powerful metaphor, true to nature or not, can be dislodged only by a more powerful metaphor. If indeed "it takes a theory to beat a theory," as the saying goes, then, as a corollary, "it takes a metaphor to beat a metaphor."

Thus, right or wrong, the Darwinian metaphor of selection is irresistably evocative. Its power to direct thought, however, may function just as well to mislead as to instruct. Because the notion of "selection" always implies the agency of a selector, from 1859 to the present many people have misinterpreted Darwin's theory so as to allow at least a beginning push by the hand of God—and, by extension, then, a divine impulse in the evolution of every species. As early as 1866, the habit of understanding the theory as one requiring supernatural agency had become so common that Alfred Russel Wallace, the theory's younger co-author, wrote to Darwin suggesting that the metaphor be jettisoned:

> I have been so repeatedly struck by the utter inability of numbers of intelligent persons to see clearly, or at all, the self-acting and necessary effects of Natural Selection, that I am led to conclude that the term itself, and your mode of illustrating it, however clear and beautiful to many of us, are yet not the best adapted to impress it on the general naturalist public.[47]

Darwin did not take the advice, and continued to employ the phrase. His stubbornness frustrated Wallace, who two years later wrote an essay entitled "Mr. Darwin's Metaphors Liable to Misconception," attempting to clarify the theory to a wider audience.[48] Neither non-scientists nor, surprisingly, biologists felt the message forcefully. Scientists and scholars continued to speak, rather sloppily, *as if* organisms were designed. As a result, in 2003 philosopher Michael Ruse lamented: "We still talk in terms appropriate to conscious intention, whether or not we believe in God.... The metaphor of design, with the organism as artifact, is at the heart of Darwinian evolutionary biology."[49] By that year, the metaphor had been shanghaied by creationists, worked into their artfully reverent euphemism "Intelligent Design," and pressed into service as an assistant to help them further muddy the intellectual waters. (I will expand upon this point in the next chapter.)

But the titular phrase was not the only Darwinian metaphor that was to cause misinterpretation and conflict. Even more confusing than "natural selection" was Darwin's use of the term "struggle for life," and assorted variations, all through the *Origin*.[50]

Economist Thomas Malthus' gloomy vision of a world always with too many children for the food available, with unceasing struggle and periodic premature deaths of large portions of many populations, was much on Darwin's mind when he wrote the *Origin*.[51] In fact, he acknowledged that his theory was "the doctrine of Malthus, applied to the whole animal and vegetable kingdom."[52] His application of the doctrine he termed "the struggle for existence," and his book is full not only of that phrase but "the struggle for life," "the war of nature," and "the great and complex battle for life."[53] He acknowledged that he used the image of warfare "in a large and metaphorical sense" so that it would include not only direct combat but "success in leaving progeny." The battle could also be indirect, although no less desperate, because "[t]wo canine animals in time of dearth may be truly said to struggle with each other which shall get food and live. But a plant on the edge of a desert is said to struggle for life against the drought."[54]

Darwin was not the only original Darwinian to advance the image of combat as a summary view of the major force driving evolution. Thomas Huxley,

Darwin's most important paladin during the nineteenth century, wrote a series of essays promoting the "gladiatorial" interpretation of natural selection.[55]

The message from evolutionary biology's founding era, then, has often seemed to be that most of life is typically nasty, brutish, and short and that only the best adapted survive to breed. It was this aspect of Darwin's message that German militarists found encouraging in their aspirations for national glory.[56] It was this aspect that Herbert Spencer found so congenial, as described in the previous chapter. And it is this aspect that has tempted many a modern biologist to indulge in Spencer-like portrayals of human society as being like nature, "red in tooth and claw."[57] For example, as Michael Ghiselin wrote in 1974, "The evolution of society fits the Darwinian Paradigm in its most individualistic form. The economy of nature is competitive from beginning to end."[58]

The inner message of the theory, then, from Darwin to Huxley to Ghiselin, seems to be that nature is an arena of "the war of all against all." Some effort may be required to realize that the "combat" description is a metaphor, that is, an interpretation nestled in language, not a feature of the scientific theory itself. All this would be of only scholastic interest absent the fact that the alleged truths of biology, translated into public discourse through metaphor, have real-world consequences. As Robert Sprinkle, a historian of both politics and biology, describes the historical pattern, "In biology's overlap with politics, explanations have moved from analogy to metaphor to policy, quite dangerously."[59]

The situation would be complicated and dangerous enough if everyone who thought about the implications of Darwinism shared the same basically Spencerian political values. The true facts of nature as revealed by evolutionary biologists, however, have provided ammunition for political thinkers at all points of the spectrum. As I reported in Chapter One, there are left-wing as well as right-wing Social Darwinists. Because so much of the political force of Darwin's theory was carried in his metaphor of eternal warlike struggle, those seeking to derive different lessons have often either attacked the metaphor itself or attempted to substitute another.

As early as 1914, the anarchist Prince Peter Kropotkin argued that in nature, cooperation is more common than competition. The prince pointed out that the first type of existential contest that Darwin described in his famous "metaphorical" passage—two wolves fighting over a kill—indeed involves direct, violent conflict. But the second type—the plant striving on the edge of the desert—is "the struggle, very often collective, against adverse circumstances."[60] Many of the species that are now extant—from ants to wolves to humans—are the species that have survived because of their penchant for intra-species cooperation. And thus, "Sociability is as much a law of nature as mutual struggle."[61]

Although his critique of the gladiatorial interpretation of Darwin's theory was trenchant, Kropotkin was unable to come up with a counter-metaphor that might have engaged the political imaginations of his audience. Indeed, while metaphors for competition are frequently encountered in the English language, metaphors for cooperation seem fewer, perhaps because they are less memorable. "Marriage," "incorporation," "friendship," "teamwork," and "family" are not quite appropriate images for the idea that Kropotkin was trying to convey, but he could not come up with a better image. Perhaps that is why he had so little influence over subsequent debates about the political implications of biological theory. Again, we see a counter-interpretation of evolution falling short for lack of rhetorical firepower.

The most important modern-era book on evolutionary theory, Richard Dawkins' *The Selfish Gene* (1976), has sometimes been misinterpreted, because of its title, as an endorsement of the nature-is-ruthless-struggle portrayal of evolution. But Dawkins' argument is mostly the opposite, an explanation of the many ways in which genes—and therefore organisms—persist through cooperation. Dawkins, a writer of considerable literary talent, finds many non-struggle metaphors to substitute for the classic Darwinian trope. Among the most vivid is the image of the racing hulls featured annually at Henley-on-Thames when two iconic English universities row competitively. "One oarsman on his own cannot win the Oxford and Cambridge boat races," he writes. "He needs eight colleagues."

> The oarsmen are genes. The rivals for each seat on the boat are alleles [the same gene with slight variations] potentially capable of occupying the same slot along the length of a chromosome.... The pool of alternative candidates is the gene pool. As far as the survival of any one body is concerned, all its genes are in the same boat.[62]

The most successful individual organism, then, is the boat that finds the best (in Darwinian terminology, the most adaptive) combination of alleles. Similarly, the most successful species will be the one that hits on the most adaptive combination of individual organisms, and that combination might very well consist of organisms that cooperate with one another against other, rival species, whether directly or indirectly.

Moreover, the best way to adapt might be not directly to challenge other species for occupation of an ecological niche but to discover or create a new niche to dominate unopposed. As Nobel Prize–winning geneticist Thomas Hunt Morgan wrote early in the twentieth century, "Evolution is not a war of all against all, but it is largely a creation of new types for the unoccupied or poorly occupied places in nature."[63]

Neither Dawkins nor Morgan was directly writing about the political implications of scientific metaphors. But either of their attempted reorientations of the dominant view of the theory of evolution is available for social thinkers who want to reconfigure the theory's implications.

None of the critics of the "struggle" metaphor is attempting to deny the fact that the natural environment destroys a large percentage of the young of each species. But those who object to the metaphor believe it is misleading because it tends to make us think that the process of evolution consists of contests to the death between individuals of the same species. By extension, given the stubborn tendency of masses of readers since 1859 to apply Darwinian metaphors to human society, it focuses attention on the competition inherent in human life as opposed to the equally important cooperation.

The employment of rival metaphors in Darwinist discourse, then, is a contest of implication. Despite the repeated insistence of some biologists that "nature contains no moral messages,"[64] human beings persist in believing that it does. And because those messages are not communicated directly and unambiguously, humans devise interpretive strategies to decode the facts of nature as presented by scientists. Thus, as Sprinkle reports, possibly misleading metaphors move toward incarnation in public policy. We might imagine metaphors struggling in an arena where those most appealing to human emotions and intellect survive while those less well adapted to our evolved preferences die out. Here, then, in a metaphor of metaphors, natural selection picks winning images for our mind's eye.

## In the Real World

Since Aristotle, serious thinkers have been trying to understand how persuasion works and why demagogues and con artists often have so much success persuading otherwise sensible human beings to take actions that are vicious, self-destructive, or plain nutty. In the last few decades, scholars have made a great deal of progress in understanding the cognitive underpinnings of human gullibility. Their major concept, which has become a staple of research in many different disciplines, is "framing."

To frame is to present information in a particular context of meaning that guides the thinking of the people who receive the information. Identical information presented to two different audiences but framed differently to each will elicit different responses. For example, cold cuts whose packaging informs potential customers that it is "90 percent fat-free" will be a bigger seller than the same meats whose package contains the message that they are "10 percent fat." In countries that ask applicants for driver's licenses to check a form giving permission to authorities to harvest their organs in the event of their accidental death, how the question is framed results in dramatic differences of compliance. In countries in which a driver must check a box to *forbid* the donation of his or her organs, agreement to be a donor is much higher than it is in those countries in which the driver must check a box to *permit* organ donation.[65]

The same principle is at work in political persuasion. In one famous experiment, various groups of subjects were shown one of two artificial television news programs about an upcoming march by the Ku Klux Klan in their town. In one of the experimental situations, the visuals were accompanied by voice-over commentary by an ersatz reporter, framing the march as a problem of the protection of free expression (should unpopular people get to have their say too?). In the other situation, identical visuals were accompanied by commentary framing the march as a problem in public order (would there be a riot?). Subjects who saw the public-order frame were considerably less likely to agree that the Klansmen should be allowed to march.[66]

A variety of other research projects have shown that to frame a story successfully, appeal, argument, and interpretation must persuade an audience to reach a framer's preferred conclusion.[67] Indeed, political contests, electoral campaigns included, can be interpreted as framing contests, with each side trying to convince undecided citizens to think about an issue using its preferred frame.[68] The term in common use among junkies of American politics, "spin doctor," refers to a person whose profession is attempting to persuade the media to adopt his or her frame of an issue or event and thus indirectly influence the way members of the public think about it.

Scholars, and the more thoughtful spin doctors, have noticed that metaphor can make a frame. As a result, political argument is often a metaphor contest. Is this candidate a pillar of the community or a fat cat? Is that one a white knight or a loose cannon? Are environmentalists tree-huggers or Earth crusaders? Is a particular grant of federal money to a local community a gift from your Uncle Sam or a greasy slice from a pork barrel? Is the ship of state sailing smoothly or about to hit the rocks?

George Lakoff, the most prolific and distinguished academic theorist of metaphor, has tried to apply his perspective to practical politics. Lakoff is a political liberal, and he is anxious to help Democrats overcome what he sees as a metaphor deficit vis-à-vis conservatives. "Do not use their language," he advises. "Their language picks out a frame—and it won't be the frame you want."[69] Right-wingers, he urged in a 2004 book, knew how to talk about values, which was why they had been winning elections. Liberals had been losing because they were insisting upon appealing rationally to citizens' self-interest. But often people did not vote their self-interest; they voted for the metaphor that activated deep emotional needs. In order to win, Democrats had to out-metaphor Republicans. Liberals would never be able to sell their metaphor to conservatives, but, if they expressed themselves persuasively, they would be able to persuade the people in the middle to endorse their metaphor.[70]

Conservatives, Lakoff said, held in their heads two metaphors that guided their issue choices and their rhetoric—the bedrock human assumptions, first, that a society was an extension of the family and, second, that the "strict father model" was the moral touchstone. The world, conservatives believed subconsciously, was a dangerous place full of evil people. It was also highly competitive. (Very much, incidentally, in line with Darwin's view of nature.) The world's competition produced winners and losers. Additionally, right and wrong were absolutely different; Lakoff, however, was not clear just how he reached this conclusion from his previous assumptions. Conservatives thus believed that what was needed in this world was a strong, strict father who could protect the family, support it in the world's competition, and teach children to distinguish good from evil.[71] As a result, conservatives supported policies permitting economic success to be rewarded and opposed policies rewarding lack of success. They supported regulation of morals according to traditional behavioral rules, and they supported strong military and police forces.

Conversely, said Lakoff, although liberals accepted at an unconscious level the same society-as-family metaphor, their image of the family in society was much different. Their idealized metaphor was of a "nurturant parent" (which was, unlike the conservative image, gender neutral). The world might be a dangerous and unjust place, but it could be made better through empathy and responsibility. It was the duty of the nurturant parent, and by extension the nurturant government, to protect and support family members. Therefore, liberals favored policies that protected the environment, workers, and consumers from abuse and exploitation.[72] The challenge for liberal candidates and spin doctors was to persuade the undecided middle to view government through *this* frame of the family.

Lakoff's advice about political metaphors was not directly relevant to the politics of evolution. As I will show in Chapter Six, there is more than one dimension to political argument. Lakoff's analysis of the American political arena, whether or not it was accurate when written, is almost entirely applicable to the *economic* dimension of conflict. Evolution belongs to the *social* dimension. There is thus no need to quibble about details here; it was Lakoff's style of discussion that remains pertinent, not the content of his advice.

The dueling-metaphors dynamic, in other words, is just as salient for "outside" politics, where pro-evolution and anti-evolution partisans debate, as it is for "inside" politics, where scientists with contending theories, hard to reconcile evidence, and contrary interpretations debate each other—and, when honestly bewildered, themselves. Darwin tried to convey his scientific ideas through metaphors of selection and struggle; in consequence, his scientific heirs state-by-state now wage a framing war with regressives eager to expel him from public school. It is my contention that the pro-evolution people are not doing well in this war (evidence supplied in Chapter Four), in large part because their use of metaphor has so often been injudicious or simply inept. Making metaphors is not their strength, and it shows. The anti-evolution people are, in general, better at fashioning persuasive rhetoric. As a result, as I will show in Chapter Four, they have the public on their side.

Biologists are perfectly willing to employ metaphors in their professional literature. In conversation with each other, they fling about such tropes as "the blitzkrieg theory,"[73] "the Swiss army knife model,"[74] "the Lazarus effect,"[75] the "house of cards" model,[76] the "snowball effect,"[77] the "Red Queen" model,[78] "arms races,"[79] and more. But in rhetorical combat with anti-evolutionists, they seem strangely reluctant to abandon the windowpane model of discourse, as though they believe that to stoop to metaphor would be to reveal themselves as non-scientific. Philosophers seem to suffer from the same inhibition. Thus, the many books by scientists and philosophers exposing the mistakes, falsities, and disinformation by creationists tend to be almost painfully precise and formal; in fact, compulsively reasonable.[80] They can be lucid and persuasive at a rational level, but they rarely or never use metaphor to speak to deep emotional needs.

The exceptions are instructive. Of the five quotations in the first paragraph of this chapter, three are from the titles of books. Titles, apparently, are free-metaphor zones. In the actual contents of those books, the authors are much more restrained in their language. Even Richard Dawkins, whose polemics against creationists are not mild-mannered, and who has shown in his own analysis of other scientists' writing that he is sensitive to the uses of metaphor, tends not to use metaphors in his own criticisms of their ideas, with the one exception noted.[81] Evidently, partisans of evolution, viewing themselves as defenders of Enlightenment rationality, are inhibited in their choice of rhetorical ammunition.

The same is not true of the anti-evolution writers. Of the quoted metaphors in the second paragraph of this chapter, none is from the title of a book. All are part of the textual message of the author. Reading them, you would not get the idea that they come from Enlightenment minds. Anti-evolutionists are pleased to portray scientific biology as a disease, a subversive influence, a Satanic conspiracy, a repulsive animal, a hoax. Their main metaphor, however, is, paradoxically, none of these bad things. Pro-evolutionists, they say, are *religious*. They worship a false god.

"In its mythological dimension," writes Phillip Johnson, "Darwinism is the story of humanity's liberation from the delusion that its destiny is controlled by a power higher than itself.... Evolution is, in short, the God we must worship."[82] James Barham agrees:

> Religion is many things, but if there is one characteristic that all religions have in common, surely it is faith ... a strong emotional attachment to an all-encompassing worldview that outstrips the available empirical evidence.[83]

Johnson and Barham are partisans of the Intelligent Design branch of creationism, but the "Young Earth" creationists (I will discuss the differences between the two in the next chapter) also argue that Darwinism is a religion.[84] Having established their premise to their own satisfaction, they can then go on to speak of scientific biology in terms of its intolerant dogma, arrogant priesthood, and unjust inquisitions.

Their argument rests on some intellectual sleight-of-hand. Creationists wish away the crucial difference between science and all religions—science must deal with natural laws, that is, laws that can be tested, studied, and refined empirically. (Terminology can become misleading, here. There is another discourse of "natural law" in philosophy, one that holds that the universe exhibits moral rules. To try to avoid causing confusion, when I refer to that tradition, I will call it "moral natural law.") The strenuous, careful, hard-nosed collection and analysis of empirical evidence since 1859 is the basis for biologists' "faith" in the theory. True, scientists must choose to rely on empirical evidence, rather than on faith, but their "faith" in the evidentiary nature of truth is not religious, as its basis can be reestablished at any time through reproducible experimentation. Faith in things seen and faith in things unseen are simply not equivalent. In the sphere of ontology, creationists choose to put their faith in magic and miracles, rejecting empirical evidence as irrelevant. In short, they eschew science and embrace religion. For them to then turn around and claim that their faith in non-empiricism is equivalent to scientists' faith in empiricism could be convincing only to people who have not been following the discussion.

But that is an argument about logical reasoning, not an argument about rhetoric. In the realm of rhetoric, the metaphorical charge that Darwinism is a religion has helped creationists convince large numbers of loose thinkers that Darwinism should, in all fairness, be compared to "other" religions in biology classes. (Creationists, not incidentally, do generally refer to "Darwinism," not "the theory of natural selection," or the "modern evolutionary synthesis," or even, generally, "evolution." Evidently, they think that by establishing biology as an "ism," they will already be partly along the road to their rhetorical destination.)

The conflation of science with religion also serves the cause of larger polemics, with the further identification of "liberalism" with "Darwinism" in the invective of the American culture war. As Ann Coulter puts it, "liberals favor one cosmology over another and demand total indoctrination into theirs."[85] Evidently, now, people who believe that natural phenomena are best studied with scientific methods are liberals—implying, ironically, that people who believe in magic are conservatives. Coulter dumps biologists along with liberals into a rhetorical stew containing people who believe that no intellectual difference exists between the sexes, that bestiality is fine among consenting animals, that criminals should not be punished for their crimes, that abortion should be performed on any girl's demand, and so on. This is a metaphorical light-year from the theory of natural selection.

Reading the many defenses of evolution by scientists and philosophers, one gets the impression that they believe implicitly that the way to get out of the rhetorical frame is by bringing ever more empirical evidence and ever more lucid rational arguments to the attention of all good citizens. No doubt that strategy has some value. But it is likely to convince only those who already think that science, not religion, should be taught to their children. To reach those who are partially suspicious of science, and as yet unconvinced about the ultimate value of scientific evolutionary theory, but remain open to persuasion, pro-science writers need to get serious about metaphor. Perhaps they should argue something like this:

The theory of evolution is a marvelous tree that blooms all year, producing fruits of many different flavors and every possible nutrient, but one that survives in only a narrow range of climate, rainfall, and fertilizer, and even then must be constantly tended, being highly vulnerable to weeds. Conversely, creationism is a zombie movie in which dead ideas keep coming back in hordes, trying to eat the brains of the living. Which, dear public, do you choose?

## Notes

1. Dawkins, Richard, *The Greatest Show on Earth: The Evidence for Evolution* (New York: Free Press, 2009).
2. Cornfield, Richard, *Architects of Eternity: The New Science of Fossils* (London: Headline Book Publishing, 2001).
3. Miller, Kenneth B., *Only a Theory: Evolution and the Battle for America's Soul* (New York: Penguin Books, 2008), 195.
4. Forrest, Barbara and Paul R. Gross, *Creationism's Trojan Horse: The Wedge of Intelligent Design* (Oxford: Oxford University Press, 2004).
5. Dawkins, Richard, *The God Delusion* (Boston: Houghton Mifflin, 2006), 113.
6. Coulter, Ann, *Godless: The Church of Liberalism* (New York: Three Rivers Press, 2007), 269.
7. Moore, J., *Should Evolution Be Taught?* (San Diego: Creation-Life Publishers, 1974), 27; quoted in Kitcher, Philip, *Abusing Science: The Case against Creationism* (Cambridge, Mass.: MIT Press, 1982), 187.
8. Poppe, Kenneth, *Reclaiming Science from Darwinism: A Clear Understanding of Creation, Evolution, and Intelligent Design* (Eugene, Ore.: Harvest House, 2006), 34, 288, 290.
9. Dembski, William A., "Introduction: The Myths of Darwinism," in William A. Dembski, ed., *Uncommon Dissent: Intellectuals Who Find Darwinism Unconvincing* (Wilmington, Del.: ISI Books, 2004), xxxiv.
10. Kaplan, Abraham, *The Conduct of Inquiry: Methodology for Behavioral Science* (Scranton, Pa.: Chandler Publishing Company, 1964), 50.
11. Frankfort-Nachmias, Chava and David Nachmias, *Research Methods in Social Science* (New York: St. Martin's Press, 1992), 27.
12. Sartori, Giovanni, "The Tower of Babel," in Giovanni Sartori, F. W. Riggs, and H. Teune, eds., *Tower of Babel: On the Definition and Analysis of Concepts in the Social Sciences* (International Studies Association, Occasional Paper no. 6, University of Pittsburgh, 1975), 67.
13. *Webster's Seventh New Collegiate Dictionary* (Springfield, Mass.: G. & C. Merriam Company, 1971), 32.
14. Kaplan, *The Conduct of Inquiry*, op. cit., 265.
15. For example, Lopez, Jose J., "Mapping Metaphors and Analogies," *American Journal of Bioethics*, 6, no. 6 (Nov./Dec. 2006), 61.

16 Hart, Roderick P. and Suzanne M. Douthton, *Modern Rhetorical Criticism*, 3rd ed. (Boston: Pearson, 2005), 2.
17 Jowett, Garth S. and Victoria O'Donnell, *Propaganda and Persuasion*, 4th ed. (Thousand Oaks, Calif.: Sage Publications, 2006), 311–313; for many other examples of the inability of scholars to explain when and why given metaphors persuade listeners, see Gibbs, Raymond W., Jr., *The Cambridge Handbook of Metaphor and Thought* (Cambridge: Cambridge University Press, 2008).
18 Quoted in Brown, Richard Harvey, "Rhetoric and the Science of History: The Debate between Evolutionism and Empiricism as a Conflict of Metaphors," *Quarterly Journal of Speech*, 72 (1986), 148–161.
19 Locke, John, *An Essay Concerning Human Understanding* (Oxford: The Clarendon Press, 1975) [originally 4th ed., 1700], 484–485, 495–496; Sprinkle, Robert Hunt, *Profession of Conscience: The Making and Meaning of Life-Sciences Liberalism* (Princeton, N.J.: Princeton University Press, 1994), 48–52.
20 Brooks, Cleanth and Robert Penn Warren, *Understanding Poetry* (New York: Holt, Rinehart, and Winston, 1938), 4.
21 Charney, Davida, "A Study in Rhetorical Reading: How Evolutionists Read 'The Spandrels of San Marco,'" in Jack Selzer, ed., *Understanding Scientific Prose* (Madison: University of Wisconsin Press, 1993), 203.
22 Lakoff, George and Mark Johnson, *Metaphors We Live By* (Chicago: University of Chicago Press, 1980), 193.
23 Gibbs, Raymond W., Jr., "Metaphor and Thought: The State of the Art," in Raymond W. Gibbs, Jr., ed., *The Cambridge Handbook of Metaphor and Thought*, op. cit., 1, 7; Hart, Roderick P., *Modern Rhetorical Criticism*, 2nd ed. (Boston: Allyn & Bacon, 1997), 146; Johnson, Mark, "Philosophy's Debt to Metaphor," in Gibbs, *The Cambridge Handbook of Metaphor and Thought*, op. cit., 39.
24 Brown, "Rhetoric and the Science of History," op. cit., 149.
25 MacCormac, Earl R., *Metaphor and Myth in Science and Religion* (Durham, N.C.: Duke University Press, 1976), 36.
26 Gibbs, Raymond, W., Jr., *The Poetics of Mind: Figurative Thought, Language, and Understanding* (Cambridge: Cambridge University Press, 1994), 172.
27 Boyd, Richard, "Metaphor and Theory Change," in Andrew Ortony, ed., *Metaphor and Thought* (London: Cambridge University Press, 1979), 356–408.
28 Kuhn, Thomas, "Metaphors in Science," in Ortony, *Metaphor and Thought*, op. cit., 418–419.
29 But for example: Gibbs, *The Cambridge Handbook of Metaphor and Thought*, op. cit.; Steen, Gerard, "The Paradox of Metaphor: Why We Need a Three-Dimensional Model of Metaphor," *Metaphor and Symbol*, 23 (2008), 213–241; Curd, Martin and J. A. Cover, eds., *Philosophy of Science: The Central Issues* (New York: W.W. Norton, 1998), section 9, "Empiricism and Scientific Realism", 1049–1289; Kitcher, Philip, *The Advancement of Science: Science without Legend, Objectivity without Illusions* (New York: Oxford University Press, 1993); Harre, Rom, *Varieties of Realism: A Rationale for the Natural Sciences* (Oxford: Basil Blackwell, 1986); Ortony, *Metaphor and Thought*, op. cit.; Hesse, Mary B., *Models and Analogies in Science* (Notre Dame, Ind.: Notre Dame Press, 1966).
30 Putnam, Hilary, "The Meaning of Meaning," in Keith Gunderson, ed., *Language, Mind, and Knowledge*, vol. 7, *Minnesota Studies in the Philosophy of Science* (Minneapolis: University of Minnesota Press, 1975).

31 Sartori, Giovanni, "Guidelines for Concept Analysis," in David Collier and John Gerring, eds., *Concepts and Methods in Social Science: The Tradition of Giovanni Sartori* (New York: Routledge, 2009), 140.
32 Grodzins, Morton, *The American System* (Chicago: Rand McNally, 1966), 8–9.
33 McCubbins, Matthew and Thomas Schwartz, "Congressional Oversight Overlooked: Police Patrols Versus Fire Alarms," *American Journal of Political Science*, 28 (1984), 165–179.
34 Dowd, Maureen, "Watch Out Below!" *New York Times*, (December 16, 2012), Opinion.
35 Kelso, John, "Fiscal Cliff Chatter Is Putting Us on Edge," *Austin American-Statesman*, (December 16, 2012), B1.
36 Stone, Deborah, *Policy Paradox: The Art of Political Decision Making*, 3rd ed. (New York: W.W. Norton, 2012), 171.
37 Dawkins, Richard, *The Blind Watchmaker: Why the Evidence of Evolution Reveals a Universe without Design* (New York: W.W. Norton, 1986, 1996), 293–296.
38 Condit, Celeste M., Benjamin R. Bates, Ryan Galloway, Sonja Brown Givens, Caroline K. Haynie, John W. Jordan, Gordon Stables, and Hollis Marshall West, "Recipes or Blueprints for Our Genes? How Contexts Selectively Activate the Multiple Meanings of Metaphors," *Quarterly Journal of Speech*, 88, no. 3 (August 2002), 303–325.
39 Ibid., 317, 320–322.
40 Darwin, Charles, *On the Origin of Species by Means of Natural Selection* (New York: Barnes and Noble Classics, 1859, 2004), 37.
41 Ibid., 35.
42 Ibid., 60.
43 Ibid., 77.
44 Gould, Stephen Jay, *The Structure of Evolutionary Theory* (Cambridge, Mass.: Harvard University Press, 2002), 621.
45 Fodor, Jerry and Massimo Piattelli-Palmarini, *What Darwin Got Wrong* (New York: Picador, 2011), 99, 114, 116.
46 Ibid., 153.
47 Wallace quoted in Robert M. Young, *Darwin's Metaphor: Nature's Place in Victorian Culture* (Cambridge: Cambridge University Press, 1985), 100.
48 Wallace, Alfred Russel, "Creation by Law" (1868), reprinted in Wallace, Alfred Russel, *Natural Selection and Tropical Nature: Essays on Descriptive and Theoretical Biology* (New York: Macmillan, 1891), 141–166.
49 Ruse, Michael, *Darwin and Design: Does Evolution Have a Purpose?* (Cambridge, Mass.: Harvard University Press, 2003), 8, 266.
50 Darwin, *Origin*, op. cit., 60, 61, 70, 74, 384.
51 Malthus, Thomas R., *Principles of Political Economy Considered with a View to Their Practical Application* (New York: Augustus M. Kelly, 1920, 1951).
52 Darwin, *Origin*, op. cit., 14.
53 Ibid., 61, 70, 74, 384.
54 Ibid., 61.
55 Huxley quoted in Gould, Stephen Jay, "Kropotkin Was No Crackpot," in *Bully for Brontosaurus: Reflections in Natural History* (New York: W.W. Norton, 1991), 328.
56 Bowler, Peter J., *Evolution: The History of an Idea*, 3rd. ed. (Berkeley: University of California Press, 2003), 305; Desmond, Adrian and James Moore, *Darwin: The Life of a Tormented Evolutionist* (New York: W.W. Norton, 1991), 542.

57 The phrase first appears in Alfred, Lord Tennyson's *In Memoriam*, written between 1833 and 1850; quoted in Bowler, *Evolution*, op. cit., 152.
58 Ghiselin, Michael T., *The Economy of Nature and the Evolution of Sex* (Berkeley: University of California Press, 1974), 247.
59 Sprinkle, Robert, "Bioethics without Analogy," in Osborne P. Wiggins and Annette C. Allen, eds., *Clinical Ethics and the Necessity of Stories: Essays in Honor of Richard M. Zaner* (Dordrecht, Netherlands: Kluwer Academic Publishers, 2013), 71–85.
60 Kropotkin, Peter, *Mutual Aid: A Factor of Evolution* (Boston: Extending Horizon Books, 1914, 1955), 60.
61 Ibid., 5, 30.
62 Dawkins, Richard, *The Selfish Gene* (Oxford: Oxford University Press, 1976, 1989), 38.
63 Morgan quoted in Loye, David, *Darwin's 2nd Revolution; Book I: Darwin and the Battle for Human Survival* (Pacific Grove, Calif.: Benjamin Franklin Press, 2010), 46.
64 Gould, Stephen Jay, *Hen's Teeth and Horse's Toes: Further Reflections on Natural History* (New York: W. W. Norton, 1983), 70.
65 Kahneman, Daniel, *Thinking, Fast and Slow* (New York: Farrar, Straus and Giroux, 2011), 88, 373.
66 Nelson, Thomas F., Rosalee A. Clawson, and Zoe M. Oxley, "Media Framing of a Civil Liberties Conflict and Its Effect on Tolerance," *American Political Science Review*, 91 (1997), 567–583.
67 Chong, Dennis, "Creating Common Frames of Reference on Political Issues," in Diana C. Mutz, Paul M. Sniderman, and Richard A. Brody, eds., *Political Persuasion and Attitude Change* (Ann Arbor: University of Michigan Press, 1996).
68 Druckman, James N., "On the Limits of Framing Effects: Who Can Frame?" *Journal of Politics*, 63, no. 4 (November 2001), 1041–1066.
69 Lakoff, George, *Don't Think of an Elephant! The Essential Guide for Progressives* (White River Junction, Vt.: Chelsea Green Publishing, 2004), 3.
70 Ibid., 21.
71 Ibid., 6–7.
72 Ibid., 11–13.
73 Raup, David M., *Extinction: Bad Genes or Bad Luck?* (New York: W.W. Norton, 1991), 81–93.
74 Carruthers, Peter and Andrew Chamberlain, "Introduction," in Carruthers and Chamberlain, eds., *Evolution and the Human Mind: Modularity, Language, and Meta-Cognition* (Cambridge: Cambridge University Press, 2000), 1–2.
75 Alroy, John, "Constant Extinction, Constrained Diversification, and Uncoordinated Stasis in North American Mammals," *Palaeo*, 127 (1996), 293.
76 Waxmen, David and Sergey Gavrilets, "20 Questions in Adaptive Dynamics," *Journal of Evolutionary Biology*, 18 (2005), 1148.
77 Orr, H. Allen, "The Population Genetics of Speciation: The Evolution of Hybrid Incompatibilities," *Genetics*, 139 (April 1995), 1805–1813, esp. 1812.
78 Dawkins, *The Blind Watchmaker*, op. cit., 183.
79 Dawkins, *The Selfish Gene*, op. cit., 250.
80 I am not going to cite every one of the very many examples that illustrate this generalization. I will content myself with offering two. By an evolutionary biologist: Coyne, Jerry A., *Why Evolution Is True* (New York: Viking, 2009). By a philosopher: Sarkar, Sahotra, *Doubting Darwin? Creationist Designs on Evolution* (Malden, Mass.: Blackwell Publishing, 2007).

81 Dawkins, Richard, *Unweaving the Rainbow: Science, Delusion and the Appetite for Wonder* (Boston: Houghton Mifflin, 1998), 180–194.
82 Johnson, Phillip E., *Darwin on Trial*, 2nd ed. (Downers Grove, Ill.: InterVarsity Press, 1993), 133, 132.
83 Barham, James, "Why I Am Not a Darwinist," in Dembski, ed., *Uncommon Dissent*, op. cit., 181.
84 Baugh, Carl, *Why Do Men Believe Evolution against All Odds?* (Bethany, Okla.: Bible Belt Publishing, 1999), 27–45.
85 Coulter, *Godless*, op. cit., 1.

# 3
# EVOLUTION AND RELIGION

One spring semester when I was in the youth of my career as a government professor, I was giving a lecture to a lower-division "Introduction to American and Texas Politics" course. The class consisted of 375 undergraduates who were required to be there by a Texas law mandating that anyone who wanted to earn a B.A. or the equivalent within the state must have instruction in the political systems of both state and nation. This particular lecture was on the subject of "ideology." It was, I explained, a system of beliefs and values, and the typical ideological system tells us how we got here, where we are going, and what is right and wrong. For purposes of illustration, I offered the members of the audience the example of Christianity. It consisted of beliefs (I offered examples), and values (more examples), and commands about right and wrong behavior, all of which were knit together into a coherent system. At the conclusion of that statement, a young woman in the second row stood up, snatched up her books, and stalked out of the lecture hall.

Thinking about this small incident afterward, I was puzzled. How could anyone, even a fervent believer, I wondered, be offended by what I had said about Christianity? How could anyone find the statement that Christianity was a system of beliefs and values that tells us how we got here, where we are going, and what is right and wrong, so insulting that they must leave the vicinity? Surely no one could doubt that what I had said about their religion was correct. So what could possibly have stirred that young woman to angry rejection?

It is not to my own credit that it took me years of chewing on the puzzle to finally be able to answer it. When I had learned enough about religion and the religious viewpoint, however, the young woman's behavior no longer seemed peculiar. After a while I came to realize that I had offended her by relativizing her faith. What I had said may have been technically accurate, but it had been contextually corrosive because it had avoided the statement that was, to her, the most important thing about Christianity—that it was true.

A religion is—or at least aspires to be—a totalized worldview. It makes commands about how we should behave that are justified with reference to an ultimate ontology. By analyzing it as if it was just another ideology in front of that class, by giving a secular interpretation to a non-secular belief system, I had implied that all religions (in fact, all ideologies) were equally true, and thus equivalently false. By going into detail about Christianity, I had made it impossible to evade the implication. No wonder that particular student became annoyed. The puzzle should have been why more of my students did not desert the lecture hall. I decided to stop using that particular example to illustrate the concept of "ideology."

In this chapter I will explore the question of whether people such as my offended student and people whose outlook is scientific can ever come to a meeting of the minds in regard to questions about the origin of species. A discussion of the relationship of religion and biology, however, requires some prefatory definitions.

As William James taught us long ago, there is not a single religious idea, temperament, or experience; there are many.[1] Even within the dominant religious tradition within the United States, the dozens of denominations, each with a slightly different take on theology, each with a slightly different prescription for human behavior, and each with a slightly different tolerance for outsiders, make any request for reference to a "Christian" point of view futile. Nevertheless, in a general sort of way, and making allowance for all the unnamed exceptions and nuances, it is possible to define "religion" as a system of beliefs, values, and behavioral prescriptions that combines ontological assumptions about supernatural agency in the universe with a system of morality that includes both moral principles and detailed rules about how humans should conduct their lives.[2]

Science, also, appears in many guises and benefits from many practices, and philosophers and historians over the past century have argued convincingly that there is no single scientific method or outlook.[3] Nevertheless, the two central organizing ideas of science are *natural law* and *empiricism*. The first is the search for non-personal, universally valid forces that explain the existence of, and interactions of, all matter and energy. The second is the relentless, systematic testing of theories about reality against measurable, observable evidence.

Religion, therefore, is centrally about personalized magic as the main explanatory concept in the universe. In contrast, science explicitly rules out personalized magic as an explanation for natural phenomena.

A final preliminary discussion concerns the concept of *secularism*. It refers to non-religious outlooks, ideas, and practices. It means just that—non-religious as opposed to anti-religious. A secular idea can be anti-religious. Atheism, for example, is simultaneously non- and anti-religious. But a secular idea need not necessarily be anti-religious. Isaac Newton, for example, propounded a series of physical laws that were entirely secular, in that they did not refer to a personalized supernatural agency. Nevertheless, Newton himself was religious, was convinced that his laws had been originally decreed by a creator, and believed that the creator might occasionally intervene in the working out of his own laws if their unassisted results seemed unsatisfactory.[4]

Whether a secular idea or theory is anti-religious depends, like beauty, on the eye of the beholder. On the one hand, an entirely spiritual consciousness, convinced that every molecule and sparrow's fall is the result of the decree of some god, will interpret any secular statement as blasphemy. On the other hand, an otherwise spiritual consciousness that accepts separate secular and religious realms—a believer such as Newton, for example—can accommodate a universe full of both religious and secular ideas. The tolerance for an idea such as the theory of natural selection, therefore, depends not so much on the theory's truth, the evidence supporting it, or the arguments on its behalf, as on the temperaments of those considering it.

## Thinking about Thinking about Evolution

We are back in the realm of implication. The first book of *Genesis* tells a story (actually, two stories, but although they differ in detail they are more or less consistent with one another in general outline) about how the universe, the Earth, all life, and human beings were created in six days by God. There are a few scholars who argue that the people who wrote *Genesis* were communicating in "mythopoeic language," and never intended the creation story to be taken literally.[5] I am not qualified to evaluate that claim. But whether it is true or not is irrelevant to a discussion of the politics of evolution. The fact is that many millions of ordinary people, not to mention a significant sample of philosophers and theologians, have regarded the creation story as a factual narrative about how literally everything got here. Because of that belief in the nature of *Genesis* as actual history, a great many Christians, now as in the past, want it to be taught as revealed truth to their children. That desire is the basis of a political issue.

Anyone who endorses the idea of "Biblical inerrancy," the conviction that every word of the *Bible* is literally true, whether having been written by God himself or written by people inspired by God, must reject any attempt to secularize the *Genesis* story. But an acceptance of the theory of natural selection as scientifically valid must imply that the story in *Genesis* cannot be literally true. Therefore, to a believer in Biblical inerrancy, the theory of natural selection has to be false. And although I am not a student of creation myths outside the Judeo-Christian tradition, the same logic must apply to them.

Anyone who can interpret the *Genesis* story as myth, allegory, or parable, however, is still free to endorse a scientific account of the history of life. If the creation story, and various narratives that go with it, such as the episode of Noah and the flood, are viewed as fictional accounts containing wisdom about human life but no actual historical grounding, then the religious implications of the theory of natural selection vanish. If *Genesis* is parable, then Darwin is not a threat to Christian belief or Christian living. The main way to undermine public hostility to the theory of natural selection, then, is to teach people to interpret *Genesis* in non-literal terms.

It appears, however, that allegorical thinking is much more difficult than it might at first appear to secular minds. Psychologists have been conducting research on the a nature of religious belief for several decades. Their conclusions have shed considerable light on the political conflicts that swirl around evolution:

1. Humans are born to believe. Children, summarizes Deborah Keleman, are "intuitive theists," that is, born with a propensity to view the world in teleological terms and to expect a "designing agent" as an explanation for natural events.[6] Religious thinking, in other words, seems to be the default option of the human mind. As Ryan Tweney observes, "There is something fundamentally easy about belief in the supernatural and there is something fundamentally hard about science.... [R]eligion is 'natural' and science is 'unnatural.'"[7]
2. Many adult minds, also, more easily accept a non-scientific worldview. As psychologist Daniel Kahneman summarizes, "Your mind is ready and even eager to identify agents, assign them personality traits and specific intentions, and view their actions as expressing individual propensities."[8] And, as Nemeroff and Rozin opine, "magical thinking is universal in adults; although the specific content is filled in by one's culture, the general forms are characteristic of the human mind."[9] This last statement must be an exaggeration, because, if it were true as stated, there would be no such thing as science. Nevertheless, it is true enough to sketch out the difficulties in getting mass publics to accept scientific theories of natural causation.
3. Aside from its propensity for personalized, magical thinking, the human mind has additional tendencies that make scientific thinking, and, in particular, evolutionary thinking, a hard sell. For example, to an evolutionary biologist, a species is not a stable entity but a "smeary continuum," in ethologist Richard

Dawkins' words, a temporary collection of genes in constant motion toward another temporary collection.[10] But ordinary human thinking is strongly *essentialist*. That is, we tend to see things, and especially living things, as individual members of clearly defined, and permanent, categories. A dog is what it is because it partakes of an almost Platonic dogness. It is essentially a dog and not an intermediate between one sort of dog and another; still less is it a frozen moment in a moving continuum of which no single moment—no species—is any more basic than any other.

Moreover, our minds have a strong tendency to observe patterns in the world around us, even if the patterns are not there. As children we have all looked at the sky and amused ourselves by picking out the portraits of duckies and bunnies and Uncle Bert that were constructed in the clouds. As adults, we see a picture of Jesus on the face of a tortilla, or the Arabic script for Allah spelled out by the beans in a stew.[11] The tendency to identify patterns, illusory or not, is so strong that psychologists have given it a name: "pareidolia." It is partly to overcome the handicap of the pareidolia tendency that scientists, and philosophers of science, have elaborated an intricate set of tests and checks to try to separate the true from the false patterns within their own perceptions.

Indeed, scientific methods can be seen as the implementation of a grand strategy to unlearn and unapply many primal habits of perception and interpretation. Nevertheless, because the human mind is, in Kahneman's phrase, such a stubborn "machine for jumping to conclusions," even the tests and checks of science often produce ambiguous conclusions. This is one of the reasons that scientific discourse is frequently full of disagreement.[12]

## Is the Theory of Evolution Compatible with Religious Belief?

The answer to the above question, then, depends on the individual viewpoint of the person answering the question. If a religious person believes in the literal truth of any given creation story, and especially the one in *Genesis*, then that person cannot also endorse modern evolutionary biology. If a religious person interprets *Genesis* as mythically powerful but historically false, however, then that person can, although not necessarily will, accept the modern evolutionary synthesis as provisionally true within the boundaries of scientific belief. Similarly, if a secular, scientifically oriented person believes that the theory of evolution implies that there is no supernatural entity anywhere in the universe, then that person must simultaneously reject all religion. If another secular person does not believe that Darwinism denies the existence of God by implication, however, then that person can tolerate or even endorse religion as a force in society.

Thus, there are not two sides to the "are they compatible" debate. There are at least four sides. As Table 3.1 illustrates, a wide variety of individual thinkers, and organizations, have sorted themselves into four quadrants based on whether they approach intellectual puzzles from a secular or faith-based perspective, and whether they believe that scientific biology and religious belief are compatible. The table is not inclusive; each quadrant could contain dozens of names and titles. I have included just enough examples in each one to illustrate the point that this is not a two-sided controversy.

Moreover, with fancy graphics the table could be made three- and perhaps four-dimensional. People and organizations who have taken positions on the "compatibility" issue, for example, could be sorted according to whether their opinions seem to originate in philosophical thinking or a desire for political accommodation. That is, commentators' positions might be categorized with reference to whether their statements are to be taken at face value or interpreted as part of a strategy to build a coalition against intellectual opponents. But I have not made this table more complex because a four-quadrant organization is sufficient to illustrate the point that there are more than two ways to answer the question of whether biology and religion are compatible.

In the upper left-hand quadrant of Table 3.1 are displayed some religious organizations that accept or even endorse the theory of natural selection, plus prominent scientists who are also religious, and prominent theologians who are

**TABLE 3.1** Acceptance and Rejection of Evolution

|  |  | Secular OR Faith-Based? | |
| --- | --- | --- | --- |
|  |  | Faith-based | Secular |
| Evolution and Religion Compatable? | Yes | Roman Catholic Church<br>United Methodist Church<br>United Church of Christ<br>Episcopal Church<br>John Haught (theologian)<br>Langdon Gilkey (theologian)<br>Francis Collins (geneticist)<br>Kenneth Miller (cell biologist) | Stephen Jay Gould (paleontologist)<br>Barry Palevitz (botanist)<br>Jerry Coyne (ecologist)<br>Michael Shermer (science journalist)<br>Michael Ruse (philosopher)<br>Jane Maienshein (biologist) |

*(Continued)*

**TABLE 3.1** Continued

|  | Secular OR Faith-Based? | |
| --- | --- | --- |
|  | Faith-based | Secular |
| No | Young Earth Creationists | Richard Dawkins (ethologist) |
|  | Intelligent Design creationists | Daniel Dennett (philosopher) |
|  | Alvin Plantinga (philosopher) | Stephen Weinberg (physicist) |
|  | Ben Stein (journalist and documentary filmmaker) | Jacob Pandian (anthropologist) |
|  |  | Victor Stenger (physicist) |

*Note*: I am here defining "science" as including methodological naturalism. Any writer who claims to be pro-science but rejects naturalism is categorized as anti-science.

*Sources*: **Upper left-hand quadrant:** for the churches, websites; Haught, John F., *Science and Faith: A New Introduction* (New York: Paulist Press, 2012), passim; Haught, *Making Sense of Evolution: Darwin, God, and the Drama of Life* (Louisville, Ky.: Westminster John Knox Press, 2010), passim; Gilkey, Langdon, *Creationism on Trial: Evolution and God at Little Rock*: (Charlottesville: University Press of Virginia, 1985), 11; Collins, Francis S., *The Language of God: A Scientist Presents Evidence for Belief* (New York: Free Press, 2006), passim; Miller, Kenneth R., *Finding Darwin's God: A Scientist's Search for Common Ground Between God and Evolution* (New York: Harper Perennial, 1999), 19, 219. **Upper right-hand quadrant:** Gould, Stephen Jay, *Rocks of Ages: Science and Religion in the Fullness of Life* (New York: Vintage, 1999), passim; Palevitz, Barry A., "Science versus Religion: A Conversation with My Students," in Paul Kurtz, ed., *Science and Religion: Are They Compatible?* (Amherst, NY: Prometheus Books, 2003), 171–180; Shermer, Michael, *Why Darwin Matters: The Case Against Intelligent Design* (New York: Henry Holt and Company, 2006), 126, 138; Ruse, Michael, *The Evolution Wars: A Guide to the Debates* (Santa Barbara, Calif.: ABC-CLIO, 2000), 282; Maienshein, Jane, "Untangling Debates about Science and Religion," in Nathaniel C. Comfort, ed., *The Panda's Black Box: Opening Up the Intelligent Design Controversy* (Baltimore: Johns Hopkins University Press, 2007), 83–108. **Lower left-hand quadrant:** Since Young Earth Creationists believe that every word of the *Bible* is literally true, they must, by definition, reject the scientific account of the history of the universe, the Earth, and life; as one among very many examples, see Faulkner, Danny, *Universe by Design: An Explanation of Cosmology and Creation* (Green Forest, Ark.: Master Books, 2004), 5–6, 96; among the Intelligent Design creationists: Johnson, Phillip E., "Evolution as Dogma: The Establishment of Naturalism," and "Creator or Blind Watchmaker?" in Robert T. Pennock, ed., *Intelligent Design Creationism and Its Critics: Philosophical, Theological, and Scientific Perspectives* (Cambridge, Mass.: MIT Press, 2001), 59–76, 435–449; Meyer, Stephen C., *Signature in the Cell: DNA and the Evidence for Intelligent Design* (New York: Harper/Collins, 2009), 435–438; Dembski, William A., *Intelligent Design: The Bridge between Science and Theology* (Downers Grove, Ill.: InterVarsity Press, 1999), 13, 119; Plantinga, Alvin, *Where the Conflict Really Lies: Science, Religion, and Naturalism* (Oxford: Oxford University Press, 2011), 83–84, 91, 96, 120–121, 283, 350; Ben Stein's attitudes are on display in his 2008 documentary, *Expelled: No Intelligence Allowed*. **Lower right-hand quadrant:** Dawkins, Richard, *The God Delusion* (Boston: Houghton-Mifflin, 2006), 54–61, and "The Great Convergence," in *A Devil's Chaplain: Reflections on Hope, Lies, Science, and Love* (Boston: Houghton-Mifflin, 2003), 146–151; Dennett, Daniel C., *Darwin's Dangerous Idea: Evolution and the Meanings of Life* (New York: Simon and Schuster, 1995), 22, 310; Weinberg, Steven, *Facing Up: Science and Its Cultural Adversaries* (Cambridge, Mass.: Harvard University Press, 2001), 230–242; Pandian, Jacob, "The Dangerous Quest for Cooperation Between Science and Religion," in Kurtz, *Science and Religion*, op. cit., 161–170; Stenger, Victor J., *Has Science Found God? The Latest Results in the Search for Purpose in the Universe* (Amherst, N.Y.: Prometheus Books, 2003), 77–95.

pro-biology. Given the noisy polemics that often surround this issue, it may be surprising to see that denominations representing a clear majority of the world's Christians accept the theory of evolution as a truthful account of the way life in general and humans in particular arrived on the scene. As historian John Hedley Brooke pointed out, theologians have often sidestepped the alleged conflict by asking some variation of the rhetorical question, "Why should evolution not be God's method of creation?"[13] Moreover, some very prominent biologists have declared that their personal faith is entirely consistent with their professional practice. Two examples are Theodore Dobzhansky, one of the founders of the "Modern Synthesis" fusing Darwinism with genetics, and Francis Collins, formerly director of the Human Genome Project and presently head of the National Institutes of Health.[14]

In the upper right-hand quadrant of the table are scientists and theologians who are clearly secular in their orientation but also maintain that there need be no conflict between scientific biology and religious faith. Many of those in this quadrant have tacitly or explicitly agreed with the argument made by paleontologist Stephen Jay Gould that science and religion occupy "Non-Overlapping Magisteria," or NOMA. Gould urged his readers to see that "the net, or magisterium, of science, covers the empirical realm. . . . The magisterium of religion extends over questions of ultimate meaning and moral value." If only people would respect the boundaries between the two, argued Gould, the supposed conflict would evaporate.[15] (Incidentally, Gould's argument has a long philosophic pedigree, originating with Immanuel Kant more than two hundred years before Gould's effort.)[16]

In the bottom right-hand column are secular scientists and other commentators who believe that scientific evolutionary theory and religious belief are incompatible. The most prominent expositor of this view during the present historical era is Richard Dawkins, whose vigorously lucid polemical style and outspoken atheism have often given the impression that he speaks for all biologists, a misunderstanding gleefully publicized by creationists. But Dawkins has spent a great deal of time attacking thinkers in all of the three other quadrants, and especially those, such as Gould, in the upper right-hand corner.[17]

In the lower left-hand quadrant are those whose perspective is religious and who believe that science and faith are incompatible. Most of these are fundamentalist Christians, although the category includes literalists of all religious traditions, including some in the United States. Most relevantly for American politics, in this quadrant reside both variations of creationism, the Young Earth Creationists and the partisans of Intelligent Design. As I will discuss shortly, those in this quadrant often deny that they are anti-science. They insist that science, as presently institutionalized, is misconceived and that its proper reorientation would permit them to be admitted to the scientific fraternity. Among the most prominent of scholars endorsing the latter position is philosopher Alvin Plantinga, whose views I will consider a bit farther on.[18]

The argument about the compatibility of biology and faith has become, in the year I write, both intense and politically relevant. But if there is one thing Darwin has taught us, it is that to understand the present nature of a complex system, we must investigate its historical development. A consideration of the evolution of thought and emotion over the past half-millennium or so, therefore, will not be a detour. It will enable us to understand the sources of the present-day disputes that have been labeled, with less exaggeration than one might think, culture wars.

## Modernism versus the Religious Temperament

There was a time when Europeans, a category that includes the ancestors of most Americans today, lived in a cultural milieu in which every aspect of life was permeated by religious meaning and religious rules. In a very rough approximation, for about the first fifteen centuries after the birth of Jesus, every thought and every behavior was saturated with Christian ideas and values, and regulated by Christian authorities. As historian Johan Huizinga put it, under those circumstances, "All thinking tends to religious interpretation of individual things; there is an enormous unfolding of religion in daily life."[19] The practical outcome of such an attitude was that social thought constantly searched for the natural law, the set of cosmic rules that combined physical reality with moral imperative, thus turning the human "is" into the sacred "ought." The ideal to which this society aspired was a wholeness of thought, feeling, and behavior in which every mundane tic of human existence was judged by the standard of the Christian ideal.

What today would be considered private life did not exist, as sexuality, among other human needs, was regulated by Christian ideas and authorities.[20] The Church did not succeed in forcing people to be chaste, of course, only in defining the boundaries of the respectable.[21] But no one questioned the claim of the Church to be arbiter of interpersonal behavior.

Thinking about economics was indistinguishable from theological moralizing, as theorists wrote about such questions as the determination of the "just price" or the sinfulness of lending money at interest.[22] The possibility that economics might have rules of operation separate from moral ideals did not occur to the European mind, or if it did, the idea was rejected as blasphemy.

Political philosophy was Christian philosophy. "Medieval political speculation," wrote Huizinga, was "imbued to the marrow with the idea of a structure of society based upon distinct orders.... [E]very one of these groupings represents a divine institution, an element of the organism of Creation emanating from the will of God."[23] All the claimants in a political dispute attempted to prove that they, literally, had God on their side.

All art was religious art. Any exceptions were underground.

Most relevantly to this chapter, there was no such separate realm of thought as science. Inquiries into matter, energy, and life were categorized under the heading "natural philosophy," and thinking about them very often started by references to passages in the *Bible*. This tendency was particularly strong in regard to the study of biology, which, of course, did not yet exist under that name. Adam and Eve and Noah's flood marked the boundaries of the possible in biological inquiry.[24]

None of this is to assert that the society of the Middle Ages was without inner conflict of many types, that individuals and groups did not produce a variety of heresies, or that Europe was not sometimes threatened by invasion from outside. Nevertheless, when the European society of the medieval centuries is contrasted with our own, it strikes us as a remarkable intellectual and moral consensus.

How and why that consensus fragmented and dissolved has been the subject of libraries of writing by historians, sociologists, philosophers, theologians, and non-academic thinkers of every stripe. There is agreement neither about causes nor about effects. Nor is there agreement about when certain crucial changes took place, so that it is not possible to identify a date or an era when the Middle Ages transitioned to the Renaissance, or when the Renaissance morphed into the Modern Age. We can identify only a few crucial historical moments that marked the arrival of a new sensibility.

When Martin Luther's Ninety-Five Theses were translated from the Latin into German in 1518, they shattered even the pretense of the unity of Christianity. When Niccolò Machiavelli published *The Prince* in 1532, he wrenched political thought out of its subjection to the philosophy of moral natural law. When Adam Smith's *Wealth of Nations* arrived in 1776, it inaugurated a way of thinking about economics that detached that new discipline from its connection to religious morality. When the United States Constitution was adopted in 1787, it established for the first time a government and a nation that were not rooted in a faith. And, running parallel to some of these, and perhaps making some of them possible, when Copernicus (1543), Kepler (1609), Galileo (1623), and Newton (1687) created the scientific revolution, they established that theories of nature must be justified, if not motivated, by non-religious arguments.

The main theme running through all these momentous events, except the first, is *secularism*. Luther had smashed the Roman church's domination of European culture, and for that and other historical reasons, that culture began a long and uneven transformation into a mode of thought that owed nothing to the theological point of view. The secular worldview is not necessarily anti-religious in intention. It can be simply indifferent to the assumptions and concerns that activate believers. But religion—any religion—is a totalized worldview. To reject a chunk of it must be to challenge all of it, if only by implication. Because the implications of secularism are inconsistent with the implications of a faith-based culture, an emerging secular ontology will inescapably generate a cultural struggle with the official ontology already in place.

In Europe, and then the Americas, the various cultural forces that proceeded from an ideological unity saturated with spiritual assumptions to a myriad of different and often inconsistent perspectives moved along separate paths at different speeds, although they were often influenced by events in other areas. Scholars of the European social development over centuries always emphasize the economic transformation from an agricultural to an industrial society, and the mass mental changes that accompany the development. As Inglehart and Welzel put it, "the publics of agrarian societies emphasize religion, national pride, obedience, and respect for authority, while the publics of industrial societies emphasize secularism, cosmopolitanism, autonomy, and rationality."[25]

Specific ideas accompanied the economic changes and had a massive, if unmeasurable, impact. Particularly Marx after 1848, Darwin after 1859, and Freud after 1900 rocked the intellectual world with powerful secular ideologies whose cumulative corrosive effect accelerated into the twentieth century. The term that I will use to characterize this overall cultural movement, consisting of the causes and consequences of a growing secularism in social thought and practice, is *modernism*, a label that I borrow from many social observers in many different disciplines.

As a new worldview fused to a set of moral assumptions, modernism has always been more a product of people who dwell in the realm of ideas than of the ordinary citizen. But ordinary people lived in the transnational society created by modernist forces and had to deal with its consequences.

In the arts, for example, the literature and painting that had been yoked to a single theme—the celebration of the Christian message—spun off into an endless number of movements and individual quests, united only by the fact that they were no longer treading the same path. As historian Peter Gay put it, by the nineteenth century, "Western civilization," meaning Europe plus its offspring in the New World and the Antipodes, "seemed to be entering a post-Christian era."[26] Many artists celebrated this liberation from faith-based assumptions and discipline by pursuing novelty as the highest goal, so that "the claim of being first and alone in the field became a central feature in the competitive modernist enterprise."[27] Karl Marx, residing by turns in Germany, the United States, and England, observed that unleashed capitalism was causing traditional economic arrangements to disintegrate, as "all that is solid melts in to air," paving the way for what he hoped would be a new, more just politics.[28]

Others were not so optimistic. In the nineteenth century, John Keats lamented the fact that the modernist philosophy would "clip an Angel's wings/Conquer all mysteries by rule and line/Empty the haunted air."[29] By the twentieth, Joseph Wood Krutch could write that the "universe revealed by science, especially the sciences of biology and psychology, is one in which the human spirit cannot find a comfortable home" and that "the most important part of our lives—our sensations, emotions, desires, and aspirations—takes place in a universe of illusions, which science can attenuate or destroy, but which it is powerless to enrich."[30]

But the opposition of some intellectuals and many ordinary citizens had, in general, little effect on the ascendancy of modernist culture. There is a variety of reasons why most citizens in the West increasingly lived under a set of cultural assumptions that a large percentage of them rejected. The main one, however, seems to be that modernism worked. It worked spectacularly in two areas.

In science, beginning with Newton, it produced understanding of reality beyond the imagination and powers over nature, beyond the fantasies of previous eras and other civilizations.

In economics, the capitalist energies unleashed by the change toward freedom of individual decision making based upon Adam Smith's arguments, when combined with the startling progress in technology that was itself partly the result of the scientific revolution, made Western nations richer and more powerful than those ever witnessed before. During the nineteenth century citizens acquired access to railroads, telephones, indoor plumbing, and indoor lighting, technological advances that made the lives of average human beings far safer, more comfortable, and more interesting than the lives of even the wealthiest of their medieval ancestors. During the twentieth, they achieved access to automobiles, antibiotics, radio and television, and, toward the end, personal computers.

The modernist trajectory of Western society, however, was neither unambiguous nor uncontested. The protests of those who wanted to preserve or go back to the medieval ontology mixed uncomfortably with the ferocious forward momentum of the secularizing tendencies of modernism through the nineteenth and much of the twentieth centuries. In particular, all Western societies contained groups of people who, as religion scholar Karen Armstrong has put it, "experienced modernism primarily as an assault."[31] Nevertheless, the fruits of modernism were in general so agreeable to humanity that they tended to intimidate potential opposition, despite the grumblings of the occasional poet or philosopher, until the 1960s.

Prior to about 1962, potentially serious conflicts between modernizing and anti-modernizing forces in Western society were stifled, short-circuited, and avoided as the almost limitless human capacity for hypocrisy smoothed transitions and justified strange bedfellows in different patterns in different countries. And the nearly as limitless capacity for fuzzy thinking and tolerance of contradiction tended to mute conflict between the incompatibles present in Western culture. As a result, despite the fact that wholehearted modernism tended to be the affiliation of a relatively small percentage of the population of Western societies, into the 1960s the twin engines of science and capitalism seemed to be the drivers of the future. Then, as is usual in history, something unexpected happened.

## Courts Go Modernist

In the United States, there was an additional dual force working for modernism, the First and Fourteenth Amendments to the Constitution. When writing their secular governing document, the framers had been careful to stipulate that no religious tests could ever be applied for federal offices. Furthermore, almost immediately after the new government came into effect, in 1791, they adopted an additional set of guarantees in the First Amendment, as part of the Bill of Rights, forbidding federal authorities to "establish" an official religion or to interfere with the free exercise of the citizens' faiths.

Because the First Amendment applied only to the activities of the central government, during the nineteenth century states or localities frequently tried to regulate the moral conduct of their residents. When public schools were established by state authorities, as they were at different rates through the century, devotional exercises, reflecting the dominant Protestant character of that society, were usually a part of the students' day.[32] Indeed, in the latter part of that century, Catholic leaders urged the creation of secular public schools as a strategy to prevent their parishioners from being subjected to the rituals and symbols of Protestantism.[33] When that strategy failed, they established the network of parochial schools that still exists.[34] Students attending public schools, therefore, generally participated in group prayer based on the King James *Bible*, students in parochial schools participated in exercises based on the Catholic version of the same book.

The Fourteenth Amendment to the Constitution, adopted in 1868, was intended to protect the civil rights of the newly freed ex-slaves, whom the ex-Confederate states were busily trying to reduce to de facto bondage. But the amendment was expansive in its rhetoric, proclaiming that no state could "abridge the privileges or immunities of citizens . . . nor deny to any person within its jurisdiction the equal protection of the laws." Beginning in the 1920s, the U.S. Supreme Court had begun using the language of this amendment to apply most of the protections of the Bill of Rights to the states, starting with freedom of expression. It took a while for the Court to get around to the "establishment of religion" clause.

In 1952, however, with the appointment of Earl Warren as Chief Justice of the United States Supreme Court, the American judicial system went modernist. In Chapter Six I will discuss the jurisprudence of the last half-century or so in detail. Here I will simply report that in 1962 the courts launched a secularizing project that still dominates the jurisprudence of evolution.

To a modernist observer, the logic of all these cases is irrefutable. As Justice Oliver Wendell Holmes wrote in another context in 1905, a constitution "is made for people of fundamentally differing views."[35] He might have expanded the principle to point out that democracy itself is a system for governing people of fundamentally differing views while avoiding the tyranny and internecine warfare that have been the two most common consequences when people of differing religious views live together. As the history of American prayers in public schools shows clearly, it is not possible to have a prayer read over an intercom system, in an institution to which children are obligated to attend, without offending the beliefs of someone. When the King James *Bible* was read in class, or when teachers led students in Protestant prayers, not only Catholics were aggrieved. As the Jewish theologian Martin Buber wrote about the generic experience of being a member of a small minority forced to undergo such an experience:

> The obligatory daily standing in the room resounding with the strange services affected me worse than an act of intolerance could have affected me. Compulsory guests, having to participate as a thing in a sacral event in which no dram of my person could or would take part.[36]

The logic applies to any member of a minority religion forced to go to public schools—to Mormons, Jehovah's Witnesses, Muslims, Hindus, Sufis, Quakers, etc. In any district, there will be a majority, perhaps Catholics in this one, Baptists in that one, or Mormons in that one. The prayers offered up will be the ones that are agreeable to the majority in that district. But in every district, there will be minority citizens who feel tyrannized by the sacral events imposed upon them. Thus, the Catholics will be the oppressed in some districts and the oppressors in others, and the same for the Mormons and the Baptists. Absent secular public schools, religious ill feelings will escalate in every district.

It is not just a theoretical possibility. In fact, in one of the court cases of religion in public school from Texas, the exact scenario I just sketched was played out for real. Baptist students in the Santa Fe (Texas, not New Mexico) high schools were leading most of the school body in Baptist prayers at football games. The parents of Catholic and Mormon students, offended, sued the school district in 1995. The U.S. Supreme Court ruled in 2000 that the prayers violated the Establishment Clause, thus saving the non-Baptist students from the sort of turmoil described in Buber's memoir.[37]

The only way that public school prayer would not be objectionable would be if every single student subscribed to the creed being propagated. But such a situation is a chimera. Permit religion into the public schools and only two outcomes are possible: The minority will submit to the tyranny of the majority, or the various creedal factions will engage in warfare. Secularism is the only workable alternative that permits social peace. This logic applies as much to the teaching of evolution as to the declaiming of prayers.

## The War on Modernism

But a large and intense minority in American society did not see the problem in this light.

We are entering an area of the discussion in which the definitions of important terms tend to be ambiguous and multiple. In my usage, a *theist* is a person who believes that God created the universe and the natural laws that govern it and that he occasionally intervenes in its operation—to answer prayers and perform miracles, for example. A *deist* is a person who believes that God created the universe and the natural laws and then permits them to work themselves out in nature and human life, without ever intervening. A *fundamentalist* is a person who believes that every word of the *Bible* is literally true. An *evangelical Christian* is one who has, at some time as an adult, established a personal relationship with Jesus and therefore feels "born again." Evangelicals and fundamentalists are both theists. Many scientists are deists. An *atheist* is a person who believes that there is no conscious, intelligent entity that created the universe and its laws. Some scientists are atheists.

Not all evangelicals are fundamentalists (former president Jimmy Carter, for example, is a non-fundamentalist evangelical), although almost all fundamentalists are evangelicals. Almost all fundamentalists, believing that *Genesis* provides a factual account of the creation, regard the theory of evolution as mistaken science at best, and Satan's propaganda at worst. Evangelicals tend to agree, although not unanimously (and here again, Carter is an example). Virtually all fundamentalists, and virtually all evangelicals, are Protestants; among scholars who study these subjects, they are generally termed *conservative Protestants*. Meanwhile, the very many denominations of non-fundamentalist, non-evangelical Protestants are usually termed *mainstream Protestants*. As organizations and as individuals, these denominations either acquiesce in or enthusiastically endorse the theory of evolution. The Roman Catholic Church officially accepts the theory of evolution as "more than just a hypothesis."[38]

Various scholars would disagree with the way I have characterized the attitude toward evolution of the major Christian traditions. I believe that it would be a waste of time to engage in the sort of definitional hair-splitting that characterizes scholarly discourse in this area, so I will merely note the differences of opinion that exist within academia, and move on.

Nevertheless, there are a few subtleties that should be noted. Not every person who rejects all or part of the theory of evolution is a fundamentalist Protestant. There are both Jews and Muslims who have gone on record as opposed.[39] Even a very small number of avowed atheist scholars have publicly rejected the scientific consensus.[40] Therefore, the general statements I am about to make, while true as descriptions of large, central tendencies of thought, are not without the occasional exception. To save time and space, I will often refer to the opposition to evolution as stemming from fundamentalists, although not every individual subscribing to the opinion falls into that category, and there are some prominent creationists who are clearly not fundamentalists.

The fundamentalist anti-evolution campaign must be understood within the context of the history of the modernist trend of American society. Darwin never claimed that he knew how life had originally arisen, and, in fact, had rhetorically conceded the possibility that the first organism might have had life "breathed into" it by a supernatural entity.[41] As of the year 2014, biologists still had not explained the origin of the first self-reproducing molecule. Given this original, and ongoing, hole in the scientific tapestry of explanation, a rational person could conclude that God might have created life and then allowed the logic of natural selection to play itself out. All it takes is an acceptance of the allegorical nature of the creation story in *Genesis*. To a person endorsing such an accommodation, a knowledge of the assumptions of modern biology is compatible with religious faith. In fact, thousands of theologians, scientists, philosophers, and ordinary citizens have endorsed this interpretation of the facts. As we shall see, however, the unprovable assertion that God might have caused the beginning of life is easy to confuse with a scientific claim that God did, in fact, do so.

But the fundamentalist mind has a "lust for certainty"[42] that refuses to accept the ambiguous half-a-loaf nature of much of modern life. Because the details of the theory of natural selection conflict with the details of *Genesis*, it must be rejected, root and branch, by fundamentalists. Moreover, because fundamentalism, at the ideological if not necessarily the personal level, tends to have delusions of persecution, "Darwinism" cannot simply be denied in isolation but must be interpreted as part of a vast modernist conspiracy against faith. Its adherents feel themselves to be both victims and targets of this conspiracy, and therefore fundamentalism becomes a "religion of rage," in Karen Armstrong's words.[43]

The fundamentalists label this conspiracy of modernism against Christianity "Secular Humanism" and sweep into its vortex all the forces that they feel are in league against goodness and truth. As the *Christian Harvest Times* editorialized in 1980:

> To understand humanism is to understand women's liberation, the ERA [Equal Rights Amendment], gay rights, children's rights, abortion, sex education ... evolution ... separation of church and state, the loss of patriotism, and many of the other problems that are tearing America apart today.[44]

And as creationist Henry Morris had written in 1978:

> [T]his question of evolution is not merely a peripheral scientific issue, but rather is nothing less than the age-long conflict between God and Satan. There are only two basic world-views. One is a God-centered view of life and meaning and purpose—the other is a creature-centered view.[45]

The fundamentalist attack on the theory of natural selection, then, is not a quibble about the interpretation of evidence. It is not a difference of viewpoint regarding how much to rely on undirected mutation as an explanation for the emergence of new body plans. It is not a disagreement about the details of evolution. It is an impassioned rejection, not just of modern biology, but of the worldview of science and the twin pillars of that worldview: secular natural law and empiricism. It is a crusade, and crusades, whether successful or not, have a way of leaving devastation in their wake.

The fundamentalist attack on the Darwinist enemy began in 1874, when Princeton theologian Charles Hodge wrote, in *What Is Darwinism?*, that it was "virtually a denial of God."[46] He concluded, however, that it was also bad science and that therefore the old-time religion was safe.[47]

By the 1920s, anti-Darwinism had strong purchase in the Southern United States, where fundamentalists had always been much more numerous than they were in the rest of the country. But for reasons embedded in American history (which I will explore in Chapter Six), the South was not close to the centers of American power. Thus, when Tennessee outlawed the teaching of evolutionary biology in public school classes, leading to the celebrated Scopes Trial of 1925, it was a state, not a federal, matter. After Scopes, the Southern states continued to deny their schoolchildren access to contemporary biological theory, and most biology textbook publishers shied away from presenting Darwinism, although many of the non-Southern states continued to teach it in the classroom.[48] Thus, although American education suffered, the issue did not again break into the national consciousness for at least four decades.

From the 1960s to the 1980s, however, three historical processes overlapped to re-energize and empower fundamentalist sentiment. First, the launch of the first Earth-orbital satellite, Sputnik, by the Soviet Union in 1957 created a national fear that free society was falling behind the Communists in science. The result was a massive federal effort to bribe and coerce the states to improve and update their science classes.[49] Suddenly, the theory of natural selection was being taught to young Americans, even in the Southern states.

Second, as already mentioned, beginning in 1962 the federal courts began imposing—or at least trying to impose—secularized education on the public schools.

Third, as I will explain in Chapter Six, political developments thrust Southern values and Southern prejudices into the center of national politics.

The cumulative effect of these historical changes was to bring the argument about evolution, and, in particular, about what to teach in public school biology classes, into the American political debate at the national, state, and local levels. In Chapters Five and Six I will analyze that debate as a political struggle. Here I will discuss and evaluate the nature of the debate on an intellectual level.

## Science versus Young Earth Creationism

The argument between the partisans of scientific biology and the partisans of creationist education has been profoundly frustrating for the scientists. They expect reason—or at least their version of reason—to prevail. They tirelessly list the evidence for the Modern Synthesis of biological theory, explain its logic, and refute the mostly nonsensical arguments against it. Then they express exasperation when, apparently, nobody but other scientists are convinced. The last four decades have witnessed a tide of Darwinian explanation and defense, all of it patient and rational, most of it well written, some of it lyrical and inspiring. The attitude of those writing books and articles in defense of modern biology is exemplified by philosopher Sahotra Sarkar's summary of his views against modern creationism: "Given the rate at which our knowledge of biology is increasing, if intellectual merit decides the outcome of the debate ... creationism has little time left."[50]

But intellectual merit is not deciding the outcome of the debate, at least not so far. During the last four decades, as I will discuss in the next chapter, American public opinion has moved, if at all, away from embracing scientific biology. Whatever the scientific, philosophical, and organizational defenders of scientific biology are doing, it is not working. That is because the opposition to the theory of natural selection is not reasonable; it is psychological and sociological, and therefore political.

Nevertheless, intellectual arguments have to be listed and considered, even if intellectual arguments are often beside the point. Creationists—those fundamentalists who specifically target scientific biology—are of several kinds, but for brevity's sake they can be divided into two camps. The first, Young Earth Creationists, are less interesting than the second, the partisans of Intelligent Design, but probably more numerous. They explicitly accept the cosmology of *Genesis* as a true statement of the origin of the universe, the Earth, and life on Earth (Young Earth Creationists usually deny that life exists anywhere else in the universe), including humans. Therefore, they insist that the Earth is less than 10,000 years old, that all the organisms that ever existed, including dinosaurs and humans, have cohabited, and that the various types of evidence of extinct creatures that have been discovered, such as fossils, can all be explained within the paradigm of Noah's flood. Some of their oft-repeated assertions are:[51]

1. The several kinds of techniques that scientists use to measure the age of the Earth, which converge on the figure of 4.6 billion years, although they are independent of one another and yet agree in most particulars, are simply wrong. The only true and reliable information about the age of the Earth is found in the *Bible*.

2. Starlight, as physicists claim, might have been traveling for billions of years before reaching Earth. Still, the Earth is only a few thousand years old, because time moves at different rates in different parts of the universe.
3. Newton's Second Law of Thermodynamics, which posits that energy steadily decays from a closed system ("entropy"), makes evolution impossible. Evolution requires organisms to accumulate energy, which, because of the unbreachable obstacle of the Second Law, can only come about with a helping hand from a supernatural force.
4. If Darwinism were true, there would be a myriad of transitional fossils preserving the tiny changes that the theory requires organisms to go through to get from one species to another. But there are no transition fossils in the rocks. What the fossil record shows is instantaneous creation. Also, all fossils were laid down in Noah's flood.
5. Modern genetics does not illustrate the slow accumulation of mutations that is required by the theory. All that genetics can show is the oscillation of already existing traits.
6. The examples often used in textbooks to illustrate evolution in action, such as the alternations of the pepper moth, *Biston betularia*, from light to dark coloration, show only "microevolution," that is, small changes within a species. They do not show "macroevolution," that is, the change from one species into another. Therefore, macroevolution does not exist.
7. Scientists have never been able to create species change in the laboratory and have never observed it in the wild. Therefore, species do not change into other species.
8. Darwinism cannot explain the origin of life. Therefore, Darwinism cannot explain the development of life after the first organism.
9. Secular Humanist scientists have engaged in an anti-Christian conspiracy for more than 150 years to suppress the truth of the *Genesis* account of biology.

The number of books and articles written by biologists, philosophers, and non-academic rationalists to refute these contentions would make a pile as high as a Diplodocus, and continues to grow.[52] It is not necessary for me to go into detail, merely to report that Young Earth Creationism's contentions have been shown to range from the preposterously mistaken (numbers 1, 3, 4, 5, 6, 7), to the irrelevant (number 8), to the paranoid (number 9). Only number 2 retains the smallest wisp of plausibility, and then only because it comes from a bona fide physicist.[53] Because it depends upon changing some of the assumptions upon which modern cosmology operates, however, we would probably be wise not to expect these suggested changes to be adopted by physicists who are not themselves Young Earth Creationists.

Despite its apparently unbreakable hold on the imagination of the masses, therefore, Young Earth Creationism offers nothing of intellectual substance for any public-spirited citizen with a respect for rational thinking. Sometimes, the only reasonable conclusion is to decide that on some questions the people are obtusely incorrect, and try to get on with things.

## Science versus Intelligent Design

That the partisans of the second branch of creationism, Intelligent Design (ID), vociferously deny that they are creationists tells us an important first fact about their strategy: It is part of an ideological marketing campaign rather than a scientific movement.[54] On the one hand, according to ID writers, biology's reliance on natural, impersonal causes to explain the development of life fails to account for the phenomenal complexity apparent in the biological world. On the other hand, it is possible to *infer* that a supernatural intelligence used magical powers to fashion all life, from bacteria to tapeworms to marigolds to humans. (Although ID writers avoid the words "magic" and "magical," that is what they mean.) Some ID theorists explicitly accept the fact that life has evolved; they maintain, however, that only an Intelligent Designer could have caused the coordinated genetic mutations that have caused the progression. Others are cagily vague about whether the Designer used directed mutations to bring about designed evolution or employed the more Biblically appropriate method of creation from scratch. But in either case, ID theorists insist that what they do is science, not religion. In fact, it is better science than the product of orthodox biology. ID is not, repeat not, a religious theory. As Stephen Meyer puts it:

> The theory of intelligent design does not claim to detect a supernatural intelligence possessing unlimited powers. Though the designing agent responsible for life may well have been an omnipotent deity, the theory of intelligent design does not claim to be able to determine that. . . . Nor can the theory of intelligent design determine whether the intelligent agent responsible for information in life acted from the natural or the 'supernatural' realm. . . . The theory of intelligent design does not claim to be able to determine the identity or any other attribute of that intelligence.[55]

All other ID writers make the same claim. They can be quite vociferous in distancing themselves from the label "creationist," given that term's grounding in *Genesis*. When William Dembski, for example, learned that two of his essays had been included in philosopher Robert Pennock's 2001 compilation *Intelligent Design Creationism and Its Critics*, he vigorously protested: "[T]here is no way I would have given my permission with that title." He would have not cooperated with Pennock's project, he explained, because creationism had become a pejorative. The strategy, therefore, should be that "as far as possible we resist being labeled." When interviewed by journalists, ID people should at all times insist that "intelligent design is not a religious doctrine about where everything came from but rather a scientific investigation into how patterns exhibited by finite arrangements of matter can signify intelligence."[56]

The claim that ID is not a religious doctrine deserves examination.

A scientist might ask that ID people come up with some sort of device to operationalize their central concept—to directly register the "design force." Such a request would have many precedents. Scientists have invented an array of technologies to measure natural forces that do not stimulate human senses—Geiger counters measure radioactive energy, EMF meters measure electromagnetic fields, and so on. This is empiricism, one of the things that distinguishes science from magic. If ID is a science, we might have expected that its practitioners would have tried very hard to come up with a "designometer" that would measure the design force. But no; apparently the Designer cannot be directly or indirectly observed. He, she, or it must be inferred. Therefore, the Designer is not only *invisible*, but *undetectable except by inference*.

The Designer, we are told, has been responsible for the development of every species over the entire 3.8 billion years of the history of life. Indeed, ID people focus such a large part of their scrutiny on the truthful accusation that the Modern Synthesis has not been able to explain the origin of life that they draw attention to the fact that the Designer was there from the beginning. Therefore, for all intents and purposes, the Designer is *immortal*.

In order to be able to manipulate the DNA of every one of the billions of species that have graced the Earth throughout that time, the Designer has had to be able to see into the core of every cell of every creature, to say nothing of being able to focus on microscopic molecules. Therefore, the Designer is *omniscient*.

Similarly, in order to cause the arrival of new species, the Designer must have had the power to manipulate those strands of DNA according to the rules of genetics. Or, if one believes that the Designer caused species to spring forth wholly formed, without the necessity of messing with genes, that theory still requires that the Designer be *omnipotent*.

Finally, there is an implication that is not part of the logic of the ID position but is usually an accompaniment to its theory. When ID theorists began to write, in the early 1990s, they were explicit about their commitment to Christianizing science. As William Dembski stated to a meeting of the National Religious Broadcasters in 2000:

> Since Darwin, we can no longer believe that a benevolent God created us in His image.... Intelligent Design opens the whole possibility of us being created in the image of a benevolent God.[57]

Since the 2005 *Dover* court decision, ID theorists have been much more cautious about proclaiming their religious intentions, and much more reluctant to identify the Designer. The most scientifically conscious of ID writers, biochemist Michael Behe, is even willing to speculate, as in this statement from a 2007 book, that "[m]aybe the designer isn't all that beneficent."[58] Nevertheless, a reading of the books and articles they have produced over two decades compels the conclusion that ID theorists all believe intensely that the Designer is *benign*.

ID theorists therefore recommend to us a Designer who is invisible, immortal, omniscient, omnipotent, and benign. They present their readers, in other words, with a Designer who has all the attributes of a Christian God and then strenuously deny the logical conclusion of their own argument. The only way to explain such a claim is to interpret it as a part of a political strategy to slip under the guards of both scientific rationalism and the First Amendment to the Constitution. In fact, ID partisans recommend this strategy in their publications, although they have become more circumspect and evasive about their true purposes as time has advanced.[59] Thus, Intelligent Design is a movement whose members consciously employ the means of deception to further their persuasive goals. As Frank Ravitch, a former advertising executive, describes the movement's procedures, "While the work of ID may never have a place in natural history museums, it should have a place in the history of great marketing strategies."[60]

The context of ID rhetoric being established, it is still useful to examine their arguments. ID theorists employ some of the same claims as the Young Earth Creationists, such as the alleged lack of transitional fossils and the inability of biologists to either create species in the laboratory or observe speciation in the wild. But these are weak arguments because they are relatively easy to refute, and ID writers do not emphasize them. Instead, they have updated, and made more sophisticated, two old arguments against Darwinism and constantly employ a philosophical argument that consists in an attack on, and a rejection of, the ontological assumptions of modern science.

The first old argument revived by ID is theologian William Paley's assertion from 1802 that it is possible to infer by looking at any entity that it was fashioned by an intelligence. Paley's example was the activity of a hypothetical person unfamiliar with modern technology, who, happening across a mechanical watch in a field next to a rock, would instantly be able to conclude that the watch had been built, whereas the rock had just happened. Paley did not specify how the human brain might differentiate the qualities of the watch from the qualities of the rock, although he seemed to be implying that the difference had something to do with complexity versus simplicity, and with the presence or absence of evident purpose. At any rate, Paley urged the reader to see that since living things were manifestly designed, there must have been a Designer.[61]

William Dembski has revived Paley's argument in a more sophisticated form. He offers a set of complicated and abstruse rules purporting to explain how design can be detected, in a watch or an organism. Unlike several other prominent ID theorists, Dembski is not an easy writer. His presentations involve a good deal of mathematics and seek to fold in information theory, physics, and various other sciences and disciplines, all rather obscurely explained. Specialists in those fields have evaluated his work, however, and found it deficient.[62] More importantly, once one has ploughed through Dembski's mathematical demonstrations, traced his use of his own eccentric definitions, and followed his argument to its conclusion, what one discovers is circular reasoning.

Disregarding Dembski's diversionary concepts, the heart of his argument is that humans can recognize the handiwork of intelligence because it combines the attributes of *complexity* and *specification*. Complexity is easy to understand. If things contain many parts that interact, a wolverine or an automobile, for example, they are complex. But the trick is to be able to decide what essential character the big weasel and the car share that permits us to infer that they were designed. The truly essential concept, therefore, is specification, and it is less easily understood.

Specified evidence is a pattern given beforehand, prior to the point at which we ask, "Is this thing designed or not?" In Dembski's words, "Specification establishes that the actualized possibility conforms to a pattern given independently of its actualization."[63] In order to illustrate the idea of specification, Dembski creates an analogy of an archer shooting an arrow into a wall. In the first try, the arrow hits the wall essentially at random. In the second, the archer first paints a ring (a bull's-eye) on the wall, then shoots the arrow into it. In the third, the archer shoots an arrow into the wall, and afterward draws a bull's-eye around it. Only on the second try, in which the arrow hit the mark previously established, would it be possible to say that the flight of the arrow had been specified.[64] As a consequence, perceiving the bull's-eye drawn on the wall, and then witnessing the arrow thudding into it, an observer would be able to infer that an intelligent agent had loosed the arrow with the intention of hitting that target.

By analogy, if we discovered Paley's watch in a field, we might intuit that its specified purpose was to mark the passage of time. (Put aside the fact that a person unfamiliar with such technology would probably not be able to imagine such a purpose.) Because the function of the machine had been specified in advance, the mind of an observer would advance to the correct conclusion that the watch had had a designer.

At this point, a critical reader anticipates that Dembski is about to deliver the punch line and explain how the human intellect can discriminate specified from non-specified complexity. How do we know that a wolverine, like a watch, and like an arrow hitting a target, is the process of conscious, purposeful construction?

And at this point, however, Dembski's publications always slough off into obscurity, irrelevancy, or a change of subject. The only way to make sense of his long excursions into the concept of specificity is to decide that he wants us to believe that since living organisms have a purpose, they are specified, and therefore designed. His whole argument is a fancy journey in a logical circle, beginning from an unacknowledged assumption, and winding back around to that assumption, dramatically portrayed as a conclusion. All of Dembski's pronouncements, therefore, which are all variations on the theme that "natural laws are in principle incapable of explaining the origin of information,"[65] are so much hoo-haw and folderol.

It was Darwin's achievement, of course, to devise a theory to explain how living things, which both are complex and seem to be accomplishing a purpose, can have been brought about through non-personal, natural processes. Darwin's accomplishment, then, is the very thing that Dembski claims cannot be accomplished. In other words, evolutionary biology long ago rendered the Paley/Dembski argument irrelevant.

The second old argument revived by ID dates back to St. George Mivart's *On the Genesis of Species*, in 1871. An outstanding zoologist of the day, he marshaled a variety of objections to Darwin's theory but emphasized one that he labeled the "incipient stages" argument.[66] Species display a variety of organs of marvelous complication, and with many of them, it is difficult to imagine how they could have come about through the lengthy, step-by-step sequence that the theory of natural selection requires. Wings, for example, are useful and complicated structures that are part of the body plans of insects, birds, and mammals. "How does evolutionary theory as understood by Darwin explain the emergence of items such as wings," he asked,

> since a small move toward a wing could hardly promote survival? . . . It is difficult, then, to believe that the Avian limb was developed in any other way than by a comparatively sudden modification of a marked and important kind.[67]

To put it in less formal language—of what use is half a wing? A creature that was beginning to develop wings would lose the value of those two limbs for grasping or running, but not yet have the advantage of flying. Would not such a creature be at a survival disadvantage, compared with its more conventionally limbed species-mates? And would it not therefore fail to survive and pass along its budding wings to the next generation? And since all proto-wings would thus cause their owners to quickly be eliminated from the genetic pool, how could fully developed wings ever evolve?

It was a powerful argument. Darwin's reply was to hypothesize the occurrence of "functional shift": Wing stubs evolved for their non-flying survival value (thermoregulation, for example) and then, at a certain point of development, became useful for flying. In the century and a half after Darwin, biologists developed and elaborated on the idea of functional shift and changed the name of the process to "exaptation." Wings, growing larger and more complex to regulate the temperature of a protobird, would suddenly be discovered to be useful for an activity for which they had not evolved—flight. Modern formal models of insects have, in fact, established the truth of the idea of small wings, useful as thermoregulators, transitioning to large wings, useful for flight. If it worked for insects, then it could have worked for birds.

Furthermore, modern studies of birds that generally live on the ground and have weakly developed wings, such as the Chukar partridge, suggest that an organism does not have to be a champion flyer to find wings useful. Birds of this species are such poor fliers that they might be described as living examples of answers to the question of the utility of half a wing. They use their wings to help them hop up onto logs and climb angled walkways. (The scientists doing these studies label the activity "wing-assisted incline running," or WAIR.) Since an organism does not have to be the "fittest" in any transcendent sense but only "fitter" than the other organisms in its competitive niche, weak wings might very well have an advantage over no wings in some environments.

In addition, observation of living flightless birds produces several illustrations of the ways that non-flying wings can have survival value. Ostriches, for example, use their wings to provide counterbalance on their turns as they run, as threat displays to intimidate predators, as sexual displays to attract mates, and as umbrellas to shade their chicks from the sun.

The answer to the question, "Of what use is half a wing?," therefore, is, "Quite a lot!"

Nevertheless, microbiologist Michael Behe, the ID theorist with the most genuine scientific credentials, has revived the "incipient stages" argument. As is befitting his profession, Behe is the creationist who goes the farthest along the road to complete acceptance of the fact of evolution. He endorses the assumption that the universe is billions of years old and finds "fairly convincing" the idea that all Earthly organisms share a common ancestor.[68] Further, he acknowledges that changes in the DNA molecule—that is, genetic mutations—are the engine of evolution.[69] Behe is thus not a fundamentalist, although everybody else in this area, including the Young Earth Creationists, borrows and endorses his ideas.

Despite his acceptance of some of the major concepts of the Modern Synthesis, Behe rejects the bedrock biological assumption that those genetic mutations are random and undirected, that they are the result neither of the needs of the organism nor of some conscious outside force. His basic position is that the molecular basis of all life, from the proteins that do the work in every cell to DNA itself, are so hideously complicated that they could not possibly have come into being through a long series of small chemical changes, each of which had to be adaptive. Essentially, he has revised Mivart's question to be, "Of what use is half a protein?" Because his answer is "No use," he concludes that all the chemicals that form the basis for living things must have come about by magic.

Behe's update of Mivart, a formulation that has made him famous and now constitutes a substantial portion of the argument of every ID writer, is "irreducible complexity." He defines it as "a single system composed of several well-matched, interacting parts that contribute to the basic function, wherein the removal of any one of the parts causes the system to effectively cease functioning."[70] An irreducibly complex system must be fully formed and composed of parts that are fully formed in order to operate. Anything other than complete perfection and it is junk. It cannot function at a lower level of efficiency; nor can it function with more or fewer parts than it has in its completeness; nor can the parts have ever had any other functions than the ones they have now.

In an inspired bit of metaphor-making, Behe chose the humble mousetrap to illustrate his idea of an irreducibly complex system. The traditional steel-spring mousetrap, he says, consists of five parts, none of which can be subtracted if the mechanism is to work, and none of which has any function outside of the mechanism. The device therefore could not possibly have evolved, step-by-step, because the constituent parts would have had no adaptive function outside the system, and the system itself would have no function without all its parts finished and properly assembled. Yet the mousetrap exists, so we know that it was intelligently designed and constructed according to a plan.[71]

To illustrate the principle of irreducible complexity in biology, Behe offers up a cornucopia of microscopic equivalents of the mousetrap—the bacterial flagellum, which rotates rapidly and propels micro-organisms through the water; the bacterial cilia, another system of propelling bacteria; the cascade of chemical interactions that constitute the clotting of blood in a human body; "vesicular transport," the system by which cells move all the right proteins to the places they are needed, at the time they are needed; the system that fits various amino acids together to form proteins; the mammalian immune system, and others. Evolutionary biologists, Behe reports, have only very rarely tried to explain how such awesomely complex chemical systems could have arisen in the necessary, step-by-step sequence. When they have tried, they have failed. The theory of natural selection, as a result, has nothing to say about the chemical basis of life:

> [I]f you search the scientific literature on evolution, and if you focus your search on the question of how molecular machines—the basis of life—developed, you find an eerie and complete silence. The complexity of life's foundation has paralyzed science's attempt to account for it.... [T]he assertion of Darwinian molecular evolution is merely bluster.[72]

Behe never states, baldly, that the reason biologists have been unable to explain the evolutionary history of various complex chemical systems is that it cannot be done, but that is what he wants us to believe. That is, he constantly moves from the *stated fact* that scientists have not been able to explain something, to the *implied principle* that they have not done it because the task is impossible—all of his examples, being irreducibly complex, cannot have evolved. But he does not show that a single one of his examples is, in principle, beyond explanation. He only presents, to an audience largely consisting of non-chemists, a series of mind-bogglingly complex microscopic systems, then asserts that they cannot be explained by science, and passes on to the conclusion that the only remaining explanation is a supernatural Designer.

Even Behe's memorable illustration of the principle, however, does not fulfill his own requirement of irreducible complexity. Biologist Kenneth Miller amuses himself in demonstrating that the little mechanism is not irreducibly complex:

> I now use a three-part mousetrap ... as a tie clip. Detach a spring from the clip, and you've got a two-part machine that works as a key chain. Glue my tie-clip to a sheet of wood, and you've got a clipboard. Attach a magnet, and you've got a refrigerator clip. ... [T]he most important part of ID's mousetrap argument—the contention that function is lost when any part of an "irreducibly complex" system is removed—fails.[73]

More importantly, it turns out that biology is not as silent as Behe would have us believe in regard to the origin of some of the molecular machines he cites. Various scientists have demonstrated the beginnings of several parts of the bacterial flagellum, for example, in other species of bacteria.[74] A precursor to the blood-clotting system in humans has been discovered in sea cucumbers, and, using that knowledge, biologists have worked out a possible adaptive sequence for the entire evolution of the system.[75] Considerable work has been done on explaining the origins of the immune system; in fact, dozens of peer-reviewed articles and at least nine books on the subject had appeared by the time of the *Dover* trial in 2004.[76] In sum, during the two decades that Behe has been insisting that, in principle, various complicated chemical reactions cannot be explained by science, scientists have made considerable progress in explaining them.

When asked about these apparent empirical refutations of his argument from principle, Behe has responded in a manner typical of both kinds of creationists. As freelance intellectual Michael Shermer puts it:

> Every time someone finds an example in nature that is simpler than Behe said it could be, Behe redefines irreducible complexity to *that* simpler level of complexity. In other words, irreducible complexity is what Behe says it is, depending on the example at hand.[77]

The final evaluation, therefore, must be that Behe's irreducible complexity argument, while superficially powerful, is upon inspection just as weak as Mivart's incipient stages argument. "Of what use is half a protein?" turns out to be no more convincing than "Of what use is half a wing?"

## A New Science?

The third common Intelligent Design argument is an attack, not just on evolutionary biology, but on the philosophical assumptions of all of modern science. Phillip Johnson, the retired Berkeley law professor who began the Intelligent Design movement in 1991 with his book *Darwin on Trial*, included not just a critique of the theory of natural selection, but a rejection of the various methodological assumptions that underlie empiricism.

The scientific enterprise does not rest on a list of cut-and-dried rules, applicable to every research project. In that sense, there is no "scientific method." But there are scientific methods, a variety of approaches to studying nature, each of which is appropriate to solving a particular problem. Despite their variety of detailed approaches, however, they all share a commitment to empiricism, a reliance on concepts that can be operationalized and measured. The observation does not, at first, have to be by the human senses. Physicists first posited the existence of electrons, then figured out ways to detect them. If they had not been able to detect them, they would not now assume that they exist. Similarly, biologists first posited the existence of genes, then invented technologies to detect them. If they had not been able to record the existence and activity of genes, the genes would not now be the conceptual basis of biology.

In either case, scientists were able to advance by adhering to what might be termed the "no miracles" rule—magical, personal, unmeasurable forces cannot be part of a scientific theory. "God did it" is not permitted as an explanation. The rule is crucial because, when a miracle is brought in to explain anything, inquiry stops. Only a reliance upon measurable concepts (empiricism), at least in principle, permits progress in scientific understanding. The philosophical underpinning of the reliance upon measurable concepts and the no-miracles rule is called *naturalism*. No naturalism, no science.

But to creationists, the implications of naturalism are evil. Their very goal is to bring back in the practice of saying "God did it" as an explanation. They refuse to be content with the idea—endorsed by very many theologians, scientists, and good citizens—that the realm of religion is one beyond the reach of science. It is all or nothing. Either we allow in miracles to explain everything in nature or we must all become atheists.

And so, Johnson devoted a chapter of his book to excoriating the "profound anti-theistic implications" of scientific biology.[78] In contrast, to theists, "the concept of a supernatural Mind in whose image we are created is the essential metaphysical basis for our confidence that the cosmos is rational and to some extent understandable.... [M]y primary goal in writing *Darwin on Trial* was to legitimate the assertion of a theistic worldview in the secular universities."[79] Naturalism does not permit supernatural explanations of natural processes, so naturalism has to go.

Johnson's point has been taken up, at some length, by philosopher Alvin Plantinga. If modern science does not permit the positing of miraculous intervention by supernatural entities as part of scientific explanations, then science must be redefined away from naturalistic assumptions. So, "What we need is a scientific account of life that isn't restricted by ... methodological naturalism," a "Christian science."[80]

Thus, Christians should insist upon a non-naturalistic science, one in which there is nothing "that conflicts with or even calls into question special divine action, including miracles."[81] Plantinga is rather vague on what a theistic science would look like, although he emphasizes that it would be just as empirically oriented as naturalistic science.[82] Again, I wish to point out that all Plantinga, or anybody else, has to do to bring special divine action into the scientific discussion of biology is to invent a designometer, a device like a Geiger counter or EMF meter that would allow him to operationalize and measure the divine power. The instant the ID people constructed such a device, ID would cease to be outside the boundaries of science and would enter the mainstream of biological discourse. But Plantinga, Johnson, and all the other ID theorists want to be accorded the title "empiricists" without actually devising anything to measure empirical reality. They want, in effect, to redefine "empiricism," which means "based on sensory observation," or the equivalent, to mean "not based on sensory observation."

The vagueness and non-explanatory nature of such a new science is well illustrated by Plantinga's efforts to come up with an example of how one might apply his methodology of scientific miracle detecting to evolutionary biology. "God might have caused the right mutations to arise in the right circumstances in such a way as to bring it about that there exist organisms of a type he intends.... [T]here is nothing in current evolutionary science to show or even suggest that God did *not* thus superintend evolution."[83]

He is right, of course. God might have shepherded evolution along a course he had planned out ahead of time. He also might have created the natural laws, then sat back to watch them work themselves out over 3.8 billion years of the history of life. Or, he might have created the natural laws, watched life struggle at the limits of single-cell existence for 3.2 billion years, then gotten bored and intervened, just once, to push life over the threshold into multicellular splendor. Or, he might have created the universe and its natural laws 13.7 billion years ago, then gotten distracted by his model train collection and lost track of time for 13 billion years, after which he looked up and was startled to discover that life had sprung up, according to those laws, without any further help. Or, he might have created the universe and its natural laws, watched the whole system work itself out, including the evolution of unicellular to multicellular life, and finally decided, about 200,000 years ago, to intervene and create *Homo sapiens*.

That is the beauty of untethering scientific explanation from operationalized, empirically-measured concepts. Any speculation, any might-have-been, becomes just as good as any other. Such speculation is easy and entertaining; it provides fodder for an unlimited number of church sermons and dormitory bull sessions; it allows philosophers to get books about science published by academic presses. The only downside is that if admitted to the list of acceptable criteria of scientific methods, it would bring the progress of human knowledge to a halt.

## Onward, through the Fog

Richard Dawkins, among others, has described Intelligent Design as "creationism in a cheap tuxedo."[84] But that is an inapt metaphor. More accurately, the ID movement should be characterized as creationism in a purloined lab coat.

To be blunt, the Intelligent Design theorists, like the Young Earth Creationists, are intellectual barbarians, seeking to reimpose a Dark Ages on human thought. But, as I shall report in the next chapter, their barbarous intentions link up well with American public opinion. Furthermore, their goals are of political relevance because their most direct policy purpose is to reintroduce the teaching of creationism—usually ID, but in some jurisdictions, possibly Young Earth—into public school biology classes. The large argument between science and creationism, therefore, is crucially important to the future of public education.

The argument over the public schools is thus at the intersection between science and politics. And the process of politics, as anyone with even casual acquaintance with it knows, is only very rarely determined on solely rational criteria. When philosopher Sahotra Sarkar expressed his hope that the controversy between science and creationism would be decided on the basis of "intellectual merit," he was expressing an ideal that we have to find inspiring.[85] But in the real world of American politics, inspiration is not enough.

Decisions in democratic politics are usually decided by the size of the coalition of disparate forces that political leaders can put together, and by the skill with which they hold together the coalition, convey the coalition's message to the media, and fit the coalition's tactics into pre-existing political interests and prejudices. The outstanding fact about pro-science people at the present time, however, is that they are not well managing their potential coalition, at least not at the level of public dispute. To go back to Table 3.1, the people listed in the top half of the table, whether religious or secular, agree that Darwinism is compatible with religion and that therefore to teach the theory of natural selection in biology classes is not to attack faith.

But the pro-evolution people in the bottom right-hand corner of the table, of whom Dawkins is the most famous and eloquent representative, insist loudly that Darwinism is in fact incompatible with religious faith. Dawkins refers to people in the upper right-hand quadrant as "the appeasement lobby" and attacks their recommendations with his usual vigorous clarity.[86] He, and the people whose views he represents, reinforce the arguments of the anti-evolutionist theists in the bottom left-hand corner, that if ordinary parents wish to protect their children from atheist propaganda, they should insist that biology classes convey religious as well as scientific messages. As Plantinga decrees the proposed doctrine, "there is a substantial segment of the population . . . whose comprehensive beliefs are indeed contradicted by the theory of evolution. . . . [T]hey have the right that public schools not teach as the settled truths beliefs that are incompatible with their comprehensive beliefs."[87]

Dawkins is right about the history of life, and Plantinga is wrong. But as a matter of political relevancy, it doesn't matter. The question is not who is right, but whose views are to be translated into public policy. The final answer to that question is yet to be decided. But Dawkins, and the others in the lower right-hand corner of the table, are not helping their side to win the political controversy. They refuse to join the broader coalition urging the compatibility of science and religion, and in so doing they not only weaken their own side, but give aid and comfort to the enemy.

There is a variety of scholars and organizations working to overcome the schism among evolutionists and fashion a broad coalition that can agree not to antagonize the sensibilities of traditional parents any more than has already been done. What has been written cannot be unwritten, but it can be supplemented with more politically sensitive statements. As philosopher Michael Ruse has urged others in this field, "given the threat that creationists pose to evolutionists of all kinds, it behooves (scientific) evolutionists especially to start thinking about working together with Christian evolutionists, rather than apart."[88]

Amen.

## Notes

1 James, William, *The Varieties of Religious Experience* (New York: New American Library, 1958); first published 1902.
2 Atran, Scott, *In Gods We Trust: The Evolutionary Landscape of Religion* (Oxford: Oxford University Press, 2002); Boyer, Pascal, *Religion Explained: The Evolutionary Origins of Religious Thought* (New York: Basic Books, 2001).
3 See the essays in Curd, Martin and J. A. Cover, *Philosophy of Science: The Central Issues* (New York: W.W. Norton, 1998).
4 Brooke, John Hedley, *Science and Religion: Some Historical Perspectives* (Cambridge: Cambridge University Press, 1991), 135–139, 144–151.

5 Hyers, Conrad, *The Meaning of Creation: Genesis and Modern Science* (Atlanta: John Knox Press, 1984), 108 and passim.
6 Keleman, Deborah, "Are Children 'Intuitive Theists'? Reasoning about Purpose and Design in Nature," *Psychological Science*, 15, no. 5 (2004), 295–301.
7 Tweney, Ryan D., "Toward a Cognitive Understanding of Science and Religion," in Roger S. Taylor and Michael Ferrari, eds., *Epistemology and Science Education: Understanding the Evolution vs. Intelligent Design Controversy* (New York: Routledge, 2011), 200.
8 Kahneman, Daniel, *Thinking, Fast and Slow* (New York: Farrar, Straus and Giroux, 2011), 76.
9 Nemeroff, Carol and Paul Rozin, "The Makings of the Magical Mind," in Karl S. Rosengren, Carl N. Johnson, and Paul L. Harris, eds., *Imagining the Impossible: Magical, Scientific, and Religious Thinking in Children* (Cambridge: Cambridge University Press, 2000), 19.
10 Dawkins, Richard, *The Blind Watchmaker: Why the Evidence of Evolution Reveals a Universe without Design* (New York: W.W. Norton, 1996), 264.
11 Haldane, David, "Believing's Seeing, Even in Chocolate," *Austin American-Statesman*, (December 22, 2006), A27.
12 Kahneman, *Thinking, Fast and Slow*, op. cit., 85, 114.
13 Brooke, John Hedley, *Science and Religion: Some Historical Perspectives* (Cambridge: Cambridge University Press, 1991), 283.
14 Collins, Francis S., *The Language of God: A Scientist Presents Evidence for Belief* (New York: Free Press, 2006), 82, 203, 230 (Collins quotes Dobzhansky on page 206).
15 Gould, Stephen Jay, *Rocks of Ages: Science and Religion in the Fullness of Life* (New York: Vintage, 1999), 6.
16 Brooke, *Science and Religion*, op. cit., 207–209.
17 Dawkins, Richard, *A Devil's Chaplain: Reflections on Hope, Lies, Science, and Love* (Boston: Houghton Mifflin, 2003), 150; Dawkins, *The God Delusion* (Boston: Houghton Mifflin, 2006), 57.
18 Plantinga, Alvin, *Where the Conflict Really Lies: Science, Religion, and Naturalism* (Oxford: Oxford University Press, 2011).
19 Huizinga, Johan, *The Waning of the Middle Ages* (New York: Doubleday Anchor Books, 1954), 151.
20 Laslett, Peter, *The World We Have Lost: England before the Industrial Age* (New York: Charles Scribner's Sons, 1965), 73, 131.
21 Tannahill, Reay, *Sex in History* (New York: Stein and Day, 1982), 256–288.
22 Dempsey, Bernard W., "Just Price in a Functional Economy," in James A. Gherity, ed., *Economic Thought: A Historical Anthology* (New York: Random House, 1965), 4–23; Aquinas, Thomas, "Of Usury," in Philip C. Newman, Arthur D. Gayer, and Milton H. Spencer, eds., *Sources in Economic Thought* (New York: W.W. Norton, 1954), 19–21 (this is a selection from *Summa Theologica*).
23 Huizinga, *The Waning of the Middle Ages*, op. cit., 57–58.
24 Brooke, *Science and Religion*, op. cit., 7–8.
25 Inglehart, Ronald and Christian Welzel, "Changing Mass Priorities: The Link between Modernization and Democracy," *Perspectives on Politics*, 8, no. 2 (June 2010), 553.
26 Gay, Peter, *Modernism: The Lure of Heresy* (New York: W.W. Norton, 2008), 27.
27 Ibid., 43.

28 Marx quoted in Berman, Marshall, *All that Is Solid Melts into Air: The Experience of Modernity* (New York: Simon and Schuster, 1982), 15.
29 Keats, John, "Lamia," in *Essential Keats* (New York: HarperCollins, 1987), 133.
30 Krutch, Joseph Wood, *The Modern Temper* (New York: Harcourt, Brace, and World, 1929, 1956), xi, 50.
31 Armstrong, Karen, *The Battle for God: A History of Fundamentalism* (New York: Random House, 2000), 4.
32 Fraser, James W., *Between Church and State: Religion and Public Education in Multicultural America* (New York: St. Martin's Press, 1999), 32–34, 87; DelFattore, Joan, *The Fourth R: Conflicts over Religion in America's Public Schools* (New Haven, Conn.: Yale University Press, 2004), 14, 21.
33 Fraser, *Between Church and State*, op. cit., 175.
34 D'Antonio, William V., Steven A. Tuch, and Josiah R. Baker, *Religion, Politics, and Polarization: How Religiopolitical Conflict Is Changing Congress and American Democracy* (Lanham, Md.: Rowman and Littlefield, 2013), 24.
35 *Lochner v. New York*, 198 U.S. 45, 25 S. Ct. 539, 49 L. Ed. 937 (1905).
36 Buber quoted in Fraser, *Between Church and State*, op. cit., 220.
37 *Santa Fe Independent School District v. Doe*, 530 U.S. 290 (2000).
38 "Vatican's Top Astronomer: Teaching of Design Wrong," *Austin American-Statesman*, (November 19, 2005), A9.
39 Edis, Taner, "A World Designed by God: Science and Creationism in Contemporary Islam," in Paul Kurtz, ed., *Science and Religion: Are They Compatible?* (Amherst, N.Y.: Prometheus Books, 2003), 117–125; Stein, Ben, *Expelled: No Intelligence Allowed*, documentary film, 2008.
40 Nagel, Thomas, *Mind and Cosmos: Why the Materialist Neo-Darwinian Conception of Nature Is Almost Certainly False* (Oxford: Oxford University Press, 2012).
41 Darwin, Charles, *On the Origin of Species by Natural Selection*, 1st ed. (New York: Barnes and Noble, 2004), 380; first published 1859.
42 Armstrong, *The Battle for God*, op. cit., 140.
43 Ibid., 216.
44 "A Special Report," *Christian Harvest Times*, June 1980, 1.
45 Morris quoted in Fraser, *Between Church and State*, op. cit., 194.
46 Ibid., 117.
47 Armstrong, *The Battle for God*, op. cit., 141–142.
48 Fraser, *Between Church and State*, op. cit., 125–126.
49 Ibid., 159.
50 Sarkar, Sahotra, *Doubting Darwin? Creationist Designs on Evolution* (Malden, Mass.: Blackwell Publishing, 2007), 130.
51 The following summary of the views of Young Earth Creationists is based on these sources: Malone, Bruce, *Censored Science: The Suppressed Evidence*, 2nd ed. (Midland, Md.: Search for the Truth Publications, 2010); Brown, Walt, *In the Beginning: Compelling Evidence for Creation and the Flood*, 8th ed. (Phoenix, Ariz.: Center for Scientific Creation, 2008); Faulkner, Danny, *Universe by Design: An Explanation of Cosmology and Creation* (Green Forest, Ark.: Master Books, 2004); Baugh, Carl E., *Why Do Men Believe Evolution against All Odds?* (Bethany, Okla.: Bible Belt Publishing, 1999); Humphreys, D. Russell, *Starlight and Time: Solving the Puzzle of Distant Starlight in a Young Universe* (Green Forest, Ark.: Master Books, 1994); Gish, Duane T., *Dinosaurs by Design*

(Green Forest, Ark.: Master Books, 1992); Gish, Duane T., *Evolution? The Fossils Say No!* 3rd ed. (San Diego: Creation-Life Publishers, 1979); Moore John N., *Questions and Answers on Creation/Evolution* (Grand Rapids, Mich.: Baker Book House, 1977); Whitcomb, John C., and Henry M. Morris, *The Genesis Flood: The Biblical Record and Its Scientific Implication* (Phillipsburg, N.J.: The Presbyterian and Reformed Publishing Company, 1961).

52  A few examples: Miller, Kenneth R., *Only a Theory: Evolution and the Battle for America's Soul* (New York: Penguin Books, 2008); Sarkar, *Doubting Darwin?* op. cit.; Futuyma, Douglas J., *Science on Trial: The Case for Evolution* (Sunderland, Mass.: Sinauer Associates, 1995); Gilkey, Langdon, *Creationism on Trial: Evolution and God at Little Rock* (Charlottesville: University Press of Virginia, 1985); Gould, Stephen Jay, *Hen's Teeth and Horse's Toes: Further Reflections on Natural History* (New York: W.W. Norton, 1983); Kitcher, Philip, *Abusing Science: The Case against Creationism* (Cambridge, Mass.: MIT Press, 1982).

53  Humphreys, *Starlight and Time*, op. cit.

54  This point is elaborated in Ravitch, Frank S., *Marketing Intelligent Design: Law and the Creationist Agenda* (Cambridge: Cambridge University Press, 2011).

55  Meyer, Stephen C., *Signature in the Cell: DNA and the Evidence for Intelligent Design* (New York: HarperCollins, 2009), 428–429.

56  Dembski, William A., "Dealing with the Backlash against Intelligent Design," in William A. Dembski, ed., *Darwin's Nemesis: Phillip Johnson and the Intelligent Design Movement* (Leicester, UK: Inter-Varsity Press, 2006), 97–98; the two essays at issue are Dembski, "Intelligent Design as a Theory of Information," 553–573, and "Who's Got the Magic?" 639–644, in Robert T. Pennock, ed., *Intelligent Design Creationism and Its Critics: Philosophical, Theological, and Scientific Perspectives* (Cambridge, Mass.: MIT Press, 2001).

57  Dembski, quoted in Forrest, Barbara, "The Wedge at Work: How Intelligent Design Creationism Is Wedging Its Way into the Cultural and Academic Mainstream," in Pennock, *Intelligent Design Creationism*, op. cit., 30.

58  Behe, Michael J., *The Edge of Evolution: The Search for the Limits of Darwinism* (New York: Free Press, 2007), 239.

59  For example: Bradley, Walter L., "Phillip Johnson and the Intelligent Design Movement," in Dembski, *Darwin's Nemesis*, op. cit., 309; see also Forrest, Barbara and Paul R. Gross, *Creationism's Trojan Horse: The Wedge of Intelligent Design* (Oxford: Oxford University Press, 2004).

60  Ravitch, *Marketing Intelligent Design*, op. cit., 209.

61  Paley, William, "Selections from *Natural Theology*," in Robert M. Baird and Stuart E. Rosenbaum, eds., *Intelligent Design: Science or Religion?—Critical Perspectives* (Amherst, N.Y.: Prometheus Books, 2007), 79–86; first published 1802.

62  Stenger, Victor J., *Has Science Found God? The Latest Results in the Search for Purpose in the Universe* (Amherst, N.Y.: Prometheus Books, 2003), 102, 104; Miller, *Only a Theory*, op. cit., 79; Gilbert, Scott F., "The Aerodynamics of Flying Carpets: Why Biologists Are Loathe to 'Teach the Controversy,'" in Nathaniel C. Comfort, ed., *The Panda's Black Box: Opening Up the Intelligent Design Controversy* (Baltimore: Johns Hopkins University Press, 2007), 49.

63  Dembski, "Intelligent Design as a Theory of Information," op. cit. 566.

64  Ibid., 560–562.

65 Dembski, William A., *Intelligent Design: The Bridge Between Science and Theology* (Downers Grove, Ill.: InterVarsity Press, 1999), 160.
66 My discussion of the incipient stages argument relies on the following sources: Gould, Stephen Jay, "Not Necessarily a Wing," in Gould, *Bully for Brontosaurus: Reflections in Natural History* (New York: W.W. Norton, 1991), 139–151; Coyne, Jerry A., *Why Evolution Is True* (New York: Viking, 2009), 57; Hanson, Thor, *Feathers: The Evolution of a Natural Miracle* (New York: Basic Books, 2011), 126–128.
67 Mivart quoted in Gould, "Not Necessarily a Wing," op. cit., 143.
68 Behe, Michael J., *Darwin's Black Box: The Biochemical Challenge to Evolution* (New York: Simon and Schuster, 1996), 5.
69 Behe, *The Edge of Evolution*, op. cit., 12.
70 Behe, *Darwin's Black Box*, op. cit., 39.
71 Ibid., 42–43, 110–111.
72 Ibid., 5, 186.
73 Miller, *Only a Theory*, op. cit., 56–57.
74 Ibid., 57–62; Coyne, *Why Evolution Is True*, op. cit., 138–139.
75 Coyne, *Why Evolution Is True*, op. cit., 139–140.
76 Miller, *Only a Theory*, op. cit., 70–74.
77 Shermer, Michael, *Why Darwin Matters: The Case against Intelligent Design* (New York: Henry Holt and Company, 2006), 67.
78 Johnson, Phillip E., *Darwin on Trial*, 2nd ed. (Downers Grove, Ill.: InterVarsity Press, 1993), 160.
79 Ibid., 164–165.
80 Plantinga, Alvin, "When Faith and Reason Clash," in Pennock, *Intelligent Design Creationism*, op. cit., 139, 141.
81 Plantinga, Alvin, *Where the Conflict Really Lies: Science, Religion, and Naturalism* (Oxford: Oxford University Press, 2011), 121–122.
82 Ibid., 268–269.
83 Ibid., 253, 46.
84 Dawkins, Richard, *The God Delusion* (Boston: Houghton Mifflin, 2006), 113.
85 Sarkar, *Doubting Darwin?*, op. cit., 130.
86 Dawkins, *God Delusion*, op. cit., 54–61, 69.
87 Plantinga, Alvin, "Creation and Evolution: A Modest Proposal," in Pennock, *Intelligent Design Creationism*, op. cit., 783.
88 Ruse, Michael, *The Evolution–Creation Struggle* (Cambridge, Mass.: Harvard University Press, 2005), 274.

# 4

# EVOLUTION AND PUBLIC OPINION

## With Tse-min Lin

Amid the tangle of beliefs, values, fantasies, and prejudices that coexist in the American public philosophy, two are relevant to the subject of evolution: democracy and science. The public affirmation of democracy goes back to the nation's birth moment in 1776, when, in the "Declaration of Independence" the new revolutionaries restated English philosopher John Locke's assertion that governments derive their legitimate authority from the consent of the governed. The public affirmation of science originated in its Constitutional founding, when its authors wrote into Article I, section 8 that Congress would have the power "to promote the progress of science and the useful arts" by granting patents and copyrights. Since 1787 there have been many who argued that the two processes of political self-government and inquiry into nature are in fact necessary to one another and that "the democratic revolution was sparked—*caused* is perhaps not too strong a word—by the scientific revolution, and that science continues to foster political freedom today."[1] While historians have pointed out that some nondemocratic societies have nurtured scientific achievement, they have added that those societies also sponsored, if only temporarily, the "organized skepticism" that is essential for science.[2] Meanwhile, if there is one thing that democratic societies are known for, it is institutionalized skepticism. The conclusion that democracy and science are natural compatriots, therefore, is an easy one to draw.

Yet democracy has always existed uneasily alongside some of its historical companions. Although the United States is famed for its population's devotion both to capitalism and to democracy, those two ideas have developed in tension all through its history.[3] And various observers since de Tocqueville in 1835 have pointed out the potential danger of "the tyranny of the majority," the extent to which large numbers of Americans have often seemed to be indifferent or hostile to the rights of people who disagree with them.[4]

So it is with evolutionary biology. Partisans of both science and democracy must deal with the reality that significant majorities of the American population, over decades of public opinion surveys, have refused to accept the truth of the theory of natural selection as an explanation for the presence of today's living organisms. Further, similarly significant majorities endorse the teaching of religious and quasi-religious theories of life's provenance in public school biology classes. Since both of these issue positions have failed to find their way into policymaking in regard to most public schools, the question arises as to how long a tension can be sustained between what the people want their children taught and what the educational establishment is willing to teach. Which will give way, biology or democracy? Or can a compromise be fashioned?

Additionally, partisans of science might argue that the problems caused by the tensions between democracy and biology could be resolved by removing the tension, that is, by educating the public so that scientific biology becomes, for the first time, the ideology of the mass and not just of the elite. After more than a century and a half of strenuous pedagogical effort, it has not happened yet, but perhaps it could be done if the right techniques were adopted.

In this chapter, I will consider both these subjects. I will first examine the evidence from public opinion surveys, including my own, about what the American (and to a lesser extent, foreign) public believes about the theory of natural selection, and the conditions under which it is acceptable to teach that theory in the public schools. Second, I will consider some of the efforts, including my own, to teach young people the truth of evolutionary theory.

## It Depends on the Question

Anyone who accepts science in general, and strives for cognitive consistency, must accept the theory of natural selection in particular. As I tried to make clear in Chapter Three, the religiously inspired critiques of Darwinism, whether of the Young Earth or Intelligent Design variety, fail abjectly. There are very few secular critiques, and these have not gained traction among thoughtful secular thinkers, and, at any rate, have had no influence over public opinion.[5] In terms of a scientific, as opposed to a faith-based, explanation for the history of life, there is no credible alternative to the Modern Synthesis of natural selection plus genetics.

Nevertheless, as Table 4.1 illustrates, among publics in every polity that has been queried, some significant percentage disbelieves. Moreover, similar significant percentages are willing to assert their belief in the myths of creation represented, among Christians and Jews, by the story in the first book of *Genesis* in the

**TABLE 4.1** Cross-National Acceptance of Evolution

Question: "Human beings, as we know them today, developed from earlier species of animals."

| Polity | Percentage of Respondents Answering "True" or "Agree" |
| --- | --- |
| Iceland | 85 |
| Denmark | 83 |
| France | 80 |
| Britain | 79 |
| Spain | 73 |
| Germany | 69 |
| Ireland | 67 |
| Poland | 59 |
| Greece | 55 |
| United States | 53 |
| Latvia | 49 |
| Texas | 35 |
| Turkey | 27 |

Question: "The earliest human beings lived at the same time as the dinosaurs."

| Polity | Percentage of Respondents Answering "True" or "Agree" |
| --- | --- |
| Sweden | 9 |
| Germany | 11 |
| Iceland | 12 |
| Denmark | 14 |
| France | 21 |
| Latvia | 27 |
| Britain | 28 |
| Spain | 29 |
| Texas | 30 |
| Italy | 32 |
| Poland | 33 |
| Turkey | 42 |

*Sources*: European data from Eurobarometer, 2005, reported in Richard Dawkins, *The Greatest Show on Earth: The Evidence for Evolution* (New York: Free Press, 2009), 433, 435; Texas data from Texas Poll, February 2010; United States from Gallup Poll, as reported by the Gallup News Service, June 11, 2007.

*Old Testament* of the *Bible*, according to which God created the universe, the Earth, and all its inhabitants in 6 days roughly 6,000 years ago.

The inference is irresistible that because belief in a recent divine creation is logically incompatible with a belief in a naturalistic development over nearly four billion years, it must be the same people endorsing the former who are denying the latter. Nevertheless, there are other possibilities. Respondents may simply be accepting the creation story and denying the fact of evolution because they have only been exposed to the first and not the second. In other words, false beliefs may be a function of deficient education, leading to an inability to recognize contradictions. Or, ignorant respondents may have prior political commitments and have been told by political authorities they trust that the incorrect explanation for present life is true and that the correct explanation is false.

Previous researchers have examined public responses to questions about creationism and evolution and come to several conclusions. First, a large chunk of the public is quite ignorant about science in general and the theory of natural selection in particular. Second, a substantial, but variable, proportion of the population rejects the Darwinian explanation of human origins outright. Third, sophisticated questions elicit a great deal of ambivalence, ignorance, and complexity in the pattern of answers. And fourth, how questions are worded has an important effect on the percentage of people falling into the various categories of response.[6]

Moreover, the enterprise of investigating these questions is hampered by the lack of consensus, among the people who study religion, as to how to conceptualize their subject matter, how to define its variables, and how to recognize the parameters of the groups they are studying. Thus, scholarship is rendered difficult by disagreement resulting from "the use of different definitional and measurement approaches, different analytic frameworks, and different populations for analysis," in the words of two prominent researchers in the field.[7]

When trying to study the effects of religious belief on non-religious attitudes, for example, previous scholars have disagreed about how to define the general concept of "religiosity," and therefore, how to sort respondents into various categories for analysis. Some scholars have classified respondents according to *membership in specific denominations*, generally identifying some Protestant denominations as "fundamentalist" while identifying different Protestant denominations, Catholics, and others as non-fundamentalist, then dividing the non-fundamentalists into either "liberal" or "moderate" categories based on their theology. Subsequently, denominations are ordered on a liberal–moderate–fundamentalist scale.[8] In other projects, researchers ask respondents about their *theological beliefs* and categorize them according to the answers. Although there is a variety of beliefs that some researchers consider crucial, the most common inquiry is about "Biblical inerrancy," or faith in the *Bible* as the literal and inspired word of God.[9] (Typically, those who aver a belief in the literal truth of the *Bible* are labeled "fundamentalists.") Other investigators ask respondents their *frequency of religious attendance*, assuming that more frequent attendance correlates strongly with more fervent faith.[10] Finally, some researchers combine a variety of measures of the respondents' beliefs and behaviors into a composite measure of an independent variable that is most commonly labeled "theological conservatism."[11] There is, then, no methodological orthodoxy among scholars in regard to how to measure various types of politically relevant religious attitudes.

But life, and research, go on amidst uncertainty. Regardless of their different methodological approaches, scholars of this subject share an assumption that religious beliefs have real-world consequences. As Leege and Kellstedt put it:

> Embedded within every religion are the basic elements of political philosophy.... Religion gives cultural expression to the main problems of existence perceived by people. It not only addresses the fundamental problems of human existence but prescribes the process of their solution and envisions the outcomes.[12]

Since the support of the theory of evolution in general, and its teaching in public schools in particular, are both religious and political issues, it is appropriate to study them together.

## The Public Speaks—or Not

The conclusion of every public opinion survey, for at least the last three decades, is that somewhat above a third of the American public rejects the theory of natural selection as an explanation for the presence of various life-forms on planet Earth, with an even larger percentage—although never quite a majority—rejecting that theory as an explanation for the presence of humans. In complementary fashion, roughly the same large minority of citizens endorses the idea that "God created humans in present form within the last 10,000 years."[13] Two questions are suggested by this research. The first is, "Who rejects evolution, and why?" The second is, "Can public opinion be changed, or, more directly, can the members of the large minority that denies the theory of natural selection be persuaded to change their minds?" I will address the first question in this section, and the second question in the next section.

The most relevant study of the first question was by Allan Mazur, and published in 2005.[14] He relied on responses to a single General Social Survey (GSS) question put to American citizens in 1993, 1994, and 2000—"How true" is the following statement: "Human beings evolved from earlier species of animals." In anticipation of my own argument, it is worth noting here that Mazur asked not about life in general, but about *Homo sapiens* in particular. I will conclude that there is a difference between these two types of questions that is relevant to the beliefs of a significant proportion of the population.

Because the GSS asked a wide range of demographic and opinion questions, Mazur was able to run correlation and regression equations in order to investigate a large variety of possible relationships. His analysis was more descriptive than theoretical; that is, he was reluctant to speculate about the possible causes of the relationships he discovered.

According to Mazur's statistics, the most important correlate of anti-(human)-evolution belief is Biblical inerrancy. Other measures of religiosity, such as membership in a fundamentalist denomination and frequency of church attendance, are independently related, but at lower levels (page 59). Self-placement on a liberal–conservative scale of political ideology has an independent effect (page 59). It would be surprising if education had no effect on beliefs about evolution, and indeed, years of schooling are positively associated (page 59) with accurate views about science. Nevertheless, even among the most educated fundamentalists, only 36 percent agree that humans evolved from earlier species. Furthermore, within both fundamentalist and non-fundamentalist groups, and controlling for education, self-described political liberals are more accepting of evolution than political moderates, and political moderates are more accepting than political conservatives (page 60). The older the respondent, the more likely to deny evolution, but age, region, race, and urban residence do not produce significant relationships (page 59).

Other surveys, without going into the multivariate analysis which was a feature of Mazur's article, suggest that simply asking one question is not sufficient to permit an understanding of the American public's views on evolution. In particular, a significantly larger percentage of Americans is willing to admit that *life* has evolved, as opposed to conceding that *humans* have evolved—about 15 percent of the population, in most surveys. Moreover, surveys that provide a "compromise" position allowing God to work through evolution between the polar positions of naturalistic evolution versus *Genesis*-type creationism elicit the support of a large chunk of the public—usually about a third of respondents.[15]

There is room, then, for some additional data and analysis of the conditions underlying beliefs about evolution.

In February 2010, I inserted four questions about evolution into the Texas Poll, which, fortuitously, is managed by my friend and colleague Daron Shaw. Thus, in addition to asking registered voters in the Lone Star State the usual questions about candidates, party identification, and so forth, the surveyors queried the 800 respondents as to their opinions on the theory of natural selection. The questions, together with the percentage of respondents endorsing each answer, were:

Q45. *Which of the following statements comes closest to your views on the origin and development of human beings?*
1. Human beings have developed over millions of years from less advanced forms of life, but God guided the process. (38 percent)

2. Human beings have developed over millions of years from less advanced forms of life, and God had no part in the process. (12 percent)
3. God created human beings pretty much in their present form about 10,000 years ago. (38 percent)
4. Don't know. (12 percent)

Q46. *Which of the following statements comes closest to your views on the origin and development of life on Earth? Life on Earth has . . .*
1. Existed in its present form since the beginning of time. (22 percent)
2. Evolved over time, entirely through "natural selection," with no guidance from God. (15 percent)
3. Evolved over time, entirely through "natural selection," but with a guiding hand from God. (53 percent)
4. Don't know. (10 percent)

Q47. *Please tell us whether you agree or disagree with the following statement: "Human beings, as we know them today, developed from earlier species of animals."*
1. Agree (35 percent)
2. Disagree (51 percent)
3. Don't know (15 percent)

Q48. *Please tell us whether you agree or disagree with the following statement: "The earliest humans lived at the same time as the dinosaurs."*
1. Agree (29 percent)
2. Disagree (42 percent)
3. Don't know (29 percent)

The results do not accurately represent all the citizens in the country. By history and geography, Texas is a Southern state and, like the other states in the region, has a population that is somewhat more religious, and, in particular, more evangelical than the residents of the rest of the country.[16] This fact may be less of a fault than it first appears, however. Firstly, no sample of a limited geographic area is representative of all human opinion. Texas may not be the United States, but the United States is not the world. Secondly, in this section I am more interested in the corollaries of belief than in the distribution of belief per se. That is, the fact that at least half the Texas population seems to disbelieve a scientific theory is of interest, but of less interest than the independent variables that help to explain the pattern of belief.

Additionally, it should be noted that these responses come from registered voters. Presumably, those who have taken the time to register are somewhat more educated and ideologically conscious than the Texas population as a whole. We might expect them, therefore, to be more intellectually sophisticated and factually informed than their unrepresented fellow-citizens.

Nevertheless, the overall distribution of opinion contains a good deal of useful information. As the percentages accompanying Question 45 illustrate, adding together the answers to the two first possible responses, 50 percent of the Texas respondents are willing to endorse either secular or theistic versions of evolution as an explanation for the existence of humans in particular. But, turning to Q46, which asks about the origins of life in general, and again summing the secular and theistic versions of the question, 68 percent of the respondents are willing to endorse the evolutionary explanation. This greater support for the idea of the evolution of life, as opposed to *Homo sapiens*, is consistent with national surveys.

This dichotomy makes no sense from the perspective of evolutionary biologists, for they treat the emergence of humans as part of the evolution of all life. Clearly, a significant percentage of the Texas respondents are looking at the questions from another perspective. Apparently, 18 percent of the respondents are neither endorsing nor rejecting a scientific perspective, but are instead answering from what might be termed an internalized political compromise, endorsing the scientific theory for life but the religious perspective for humans.

The same sort of yearning for compromise appears when the responses to Q47 are contrasted with the responses to Q45. Q47 offers a stark, dichotomous choice to the respondents—did human beings evolve from animals, yes or no. With only those two alternatives, an actual majority chooses the *Genesis* explanation. But when, in Q45, Texans are offered three choices—yes, no, and yes but with a guiding hand from God—the proportion of respondents rejecting evolution falls by about 15 percent and the compromise choice ties the *Genesis* choice in popularity.

Meanwhile, there is a hard core of respondents—from about a fifth to about two-fifths of the population, depending on the way the question is worded—who endorse the literal religious explanation and reject the scientific explanation. And perhaps most surprisingly, the secular scientific explanation attracts the support of, at most, 15 percent of the Texas population. Along with other surveys taken over a generation, this one emphasizes the extent to which the endorsement of scientific biology has only a very tenuous hold on the imagination of the public.

Multivariate analysis provides more insight into the dynamics of these opinions. What follows draws on research I conducted in 2010 for an American Political Science Association paper with my co-author, Brian Roberts.[17]

For the two questions with three substantive answers, we first made the reasonable assumption that there is a naturally ordered progression through the question responses—scientific, compromise, and religious. We then estimated logistical models with separate dependent variables based on the secular ordering of the substantive responses. Our hypothesis was that the more religious the respondents tended to be, the more they would endorse the *Genesis* theory of the origin of humans (and vice versa), and the less religious, the more they would reject it (and vice versa).

Among the independent variables, we used two questions from the Texas poll that measured religious commitment, one that asked respondents to express the importance of religion in their daily lives ("religiosity"), the other asking whether the respondents considered themselves evangelical Christians. As Table 4.2 illustrates, both of these variables are of great help in explaining who accepts or rejects the scientific understanding of evolution. On the left side of the table the secular ("Model A"), compromise ("Model B"), and creationist ("Model C") responses are arrayed as dependent variables, while down the left border of the table are arrayed the independent variables from the Texas Poll. The statistics in Model C emphasize the strong relationships between positioning oneself as an evangelical or as highly religious and the tendency to endorse the theory that God created humans within the last 10,000 years. Examining Model B, we see that evangelicals

**TABLE 4.2** Statistical Relationships for Secular, Compromise, and Creationist Survey Responses

|  | Origins of Man[1] | | | Origins of Life | | |
| --- | --- | --- | --- | --- | --- | --- |
|  | Model A | Model B | Model C | Model A | Model B | Model C |
| Constant | 2.19** | −0.69 | −4.24*** | 3.92*** | −0.96* | −5.80*** |
|  | (−0.98) | (0.58) | (0.76) | (0.98) | (0.57) | (1.15) |
| Male | 0.61* | 0.10 | −0.25 | 0.78** | 0.15 | −0.35 |
|  | (0.34) | (0.19) | (0.21) | (0.33) | (0.18) | (0.23) |
| Black | −0.46 | −0.60** | 0.18 | −0.70 | −0.34 | 0.30 |
|  | (0.52) | (0.29) | (0.34) | (0.51) | (0.28) | (0.34) |
| Hispanic | −0.10 | 0.08 | −0.32 | −0.41 | −0.16 | 0.31 |
|  | (0.47) | (0.24) | (0.26) | (0.43) | (0.24) | (0.31) |
| Evangelical | −0.41 | −1.01*** | 0.82*** | −0.97* | −0.55*** | 0.74** |
|  | (0.56) | (0.21) | (0.21) | (0.57) | (0.21) | (0.24) |
| Relig. important | −1.32*** | 0.38*** | 0.77*** | −1.55*** | 0.52*** | 0.80*** |
|  | (0.19) | (0.12) | (0.16) | (0.19) | (0.12) | (0.23) |
| Non-religious | 1.14*** | −0.59** | −0.74 | 0.60* | −0.62** | −0.68 |
|  | (0.35) | (0.30) | (0.45) | (0.36) | (0.28) | (0.69) |
| Age | 0.02 | −0.01 | −0.01 | 0.01 | −.01** | 0.003 |
|  | (0.01) | (0.01) | (0.01) | (0.01) | (0.006) | (.01) |
| Education | 0.17 | 0.20*** | −0.25*** | 0.11 | −0.10 | −0.11 |
|  | (0.13) | (0.07) | (0.08) | (0.11) | (0.07) | (0.10) |
| Income | −0.03 | 0.02 | −0.01 | −0.02 | 0.02 | −0.02 |
|  | (0.05) | (0.03) | (0.03) | (0.05) | (0.03) | (0.03) |
| Conservative | −0.55*** | −0.19*** | 0.46*** | −0.57*** | −0.03 | 0.36*** |
|  | (0.12) | (0.06) | (0.08) | (0.11) | (0.06) | (0.084) |
| Number of obs | 703 | 703 | 703 | 703 | 703 | 703 |
| F–stat | 10.01 | 6.31 | 11.96 | 12.96 | 4.52 | 6.55 |
| Prob > F | 0.000 | 0.000 | 0.000 | 0.000 | 0.000 | 0.000 |

*Note:* [1] Standard errors in parentheses. Statistical significance: *** (<.01), ** (<.05), * (<.10).

reject the compromise position but that religious Texans in general endorse it. And then, observing Model A, it is clear that both evangelicals and the average religious person reject it. Again, then, we see that while a large minority of Texans are resolutely creationist, a significant percentage, while they abhor the entirely secular explanation for the presence of life, are willing to embrace an evolutionary explanation of that life if the explanation is resting on a theistic base.

The relationships using Q46 (evolution of life) as a dependent variable, displayed on the right side of Table 4.2, tell substantially the same story. Again, evangelicals rebuff both the secular and compromise positions, and again, the religious in general, while being repulsed by the secular position, are willing to endorse the compromise position.

Table 4.2 also makes clear that self-described political conservatives resemble evangelicals more than they resemble typical Christians. Like the evangelicals, they reject not only the secular response to Q45 ("humans"), but the compromise choice as well. The coefficients are smaller than they are for evangelicals, perhaps meaning that political conservatives are, as a group, less vociferous in their rejection of a scientific explanation for life and humans, but the relationship is still strong and statistically significant.

The pattern is the same, except in the case of the compromise position in Q46 ("life"). The responses of Texas political conservatives have no significant relationship to the compromise position. As with Q45, however, conservatives reject the secular stance and embrace the creationist.

Political conservatism and religiously motivated rejection of the scientific theory of life are, then, clearly related to one another. In Chapter Six, I will discuss the way this lumping of all "conservatives" into one category misses some important distinctions. Since I must use the one variable here, however, I will proceed but caution that the generalizations will be refined in Chapter Six.

For decades, scholars have been doing research on the personality traits associated with political conservatism—their origin and their expression. The conclusion of this large amount of research is that "the common basis for all the various components of the conservative attitude syndrome is a *generalized susceptibility to experiencing threat or anxiety in the face of uncertainty*,"[18] which results in, among other consequences, "religious orthodoxy,"[19] "superstition, religious dogmatism,"[20] and "mental rigidity and close-mindedness."[21]

In other words, political conservatives, like religious evangelicals (and especially the fundamentalist subset of evangelicals), are motivated both to perceive the theory of natural selection as a personal threat and to react to that threat by rejecting evenhanded evaluation of evidence and embracing dogmatic ideological authority. Not only religious fundamentalists, then, but political conservatives seem susceptible to the same sort of anti-modernism reaction, the historical development of which I described in Chapter Three.

The political science literature on public opinion over the past half-century would lead us to expect that more education would be strongly associated with acceptance of a scientific theory. As Table 4.2 shows, however, this expectation is only partially borne out. Texans with more education do clearly reject the *Genesis* story as an explanation for the origin of humans, but their responses have no significant relationship to the positions on the evolution of life. As I will demonstrate in the next section of this chapter, education does not necessarily translate into an endorsement of scientific biology. As a result, we should be cautious in interpreting the statistical relationships in Q45 and Q46 as illustrating a causal connection between education and endorsement of any particular worldview. The question remains open and suggests a path for future research.

In Table 4.3, I present some intercorrelations of answers to the four questions on the Texas survey. Some of my readers here will be quite familiar with correlation coefficients. Others may be encountering them for the first time. To those who are unfamiliar, let me briefly explain that the coefficients express a measurement of covariation between two variables—the extent to which two types of behavior or opinion are found together. The statistical theory underlying correlations need not concern us. Suffice it to say that correlations run from −1 (the two variables are never found together) through 0.00 (there is no association between the two variables, any coexistence being random) to +1 (the two variables are always found together).

At the top of the table are relationships between expressed opinions supporting evolution, and at the bottom are relationships between expressed opinions denying it. The correlations suggest the same ambivalence and willingness to embrace ideological compromise as those implied by Table 4.2. At the top of the table, the correlation of the opinions of those endorsing evolution of life and evolution of humans is a very strong 0.84, just what we would expect of people who embrace the theory comprehensively. The correlation between those and the fourth question, about whether humans and dinosaurs lived together, however, is noticeably less. Similarly, the intercorrelations among the opinions denying evolution are relatively strong, although considerably weaker than those for the people who endorse evolution. Notice, however, that the relationship between the two questions inquiring about human origins correlate more strongly with each other than either does with the question inquiring about the origin of life in general. This is further evidence that the wording of the question matters.

Nevertheless, the minuscule correlations between evolution denial and belief in human/dinosaur cohabitation are puzzling. Those who embrace the idea that the Earth is less than 10,000 years old and do not deny the evidence of fossils must also believe that the intelligent bipeds and the giant reptiles must have walked the planet at the same time. Indeed, as I reported in Chapter Three, this is exactly what books by Young Earth Creationists teach. Yet the correlations in the bottom

**TABLE 4.3** Intercorrelations of Opinions Supporting and Rejecting Evolution in Texas

*Supporting*

| Question | 1 | 2 | 3 | 4 |
|---|---|---|---|---|
| 1 | 1.00 | | | |
| 2 | 0.84 | 1.00 | | |
| 3 | 0.41 | 0.46 | 1.00 | |
| 4 | 0.18 | 0.17 | 0.24 | 1.00 |

All pairwise correlations are statistically significant at the level of $p < .001$.
Question 1: Human beings have developed over millions of years from less advanced forms of life, and God has had no part in the process.
Question 2: Life on Earth evolved over time, entirely through "natural selection," with no guidance from God.
Question 3: Agree that human beings, as we know them today, developed from earlier species of animals.
Question 4: Disagree that the earliest humans lived at the time of the dinosaurs.

*Rejecting*

| Questions | 1 | 2 | 3 | 4 |
|---|---|---|---|---|
| 1 | 1.00 | | | |
| 2 | 0.46 | 1.00 | | |
| 3 | 0.62 | 0.39 | 1.00 | |
| 4 | 0.13 | 0.15 | 0.10 | 1.00 |

Question 1: God created human beings pretty much in their present form about 10,000 years ago.
Question 2: Life on Earth has existed in its present form since the beginning of time.
Question 3: Disagree that human beings, as we know them today, developed from earlier species of animals.
Question 4: Agree that the earliest humans lived at the same time as the dinosaurs.

half of Table 4.3 suggest that there is no relationship between the professed beliefs of many Texans and the logical consequence of those beliefs.

I suspect that the puzzling lack of relationship in the table exists because the fourth question is of a somewhat different kind than the first three. In effect, they ask for opinions about a theory. The "dinosaurs" question, in contrast, taps a reservoir of factual knowledge. Almost everyone has an opinion; few are conversant with empirical reality. The fact that 29 percent of the Texas respondents answered

"Don't know" to this question strengthens the interpretation that it simply measures a large amount of ignorance in the population, whether people endorse the creationist, compromise, or scientific position.

The result of all this examination of statistical relationships is a list of several bottom-line generalizations about the public opinion of evolution:

1. Only a rather small percentage of the American (and Texas) population endorses the validity of the scientific theory of evolution in its correct, entirely secular form. But a significantly larger percentage of respondents will agree to accept the truth of evolution if they are allowed to insert the caveat that "God guided the process." This large minority in the Texas sample seems to conform to Putnam and Campbell's description of most Americans as embracing a "faith without fanaticism," which concedes that "there are many ways to get to heaven."[22] Or, it may reveal a section of the population that is reluctant to choose between a thoroughgoing naturalistic attitude toward life and a consistently spiritual attitude. Although they would not put it this way, they may prefer muddled compromise to clearly articulated conflict.

    Members of the public at large, in other words, may find themselves nestled in the top left-hand quadrant of Table 3.1 from the previous chapter, without having actually thought themselves there. A large percentage of ordinary people may be religious in their personal lives but are willing to make room for quasi-scientific biology at the level of public philosophy.
2. Similarly, questions that ask respondents about their views on the evolution of life in general or of human beings in particular elicit dramatically different distributions of opinion. Many are willing to acquiesce in some version of the naturalistic development of life, but few are willing to accept that their own species is included in the larger pattern.
3. When questioning respondents about their views on evolution, items that draw forth opinions are not equivalent to questions that ask for factual knowledge. Ignorant people are generally willing to offer theoretical opinions, but often reluctant to claim empirical expertise.
4. Political ideology has a strong independent effect on the tendency to accept or deny the theory of evolution. It would seem that both political and religious conservatism are similarly rooted in human personality.

## Teaching Darwin—or Not

The fact of the distribution of public opinion about evolution is, if viewed in isolation, simply academic. But it becomes relevant to real policy decisions when the questions pertain to the content of public school biology classes. With the usual caveats about question wording, substantial majorities of the American public have, for at least the last generation, endorsed the teaching of both creationism and natural selection, in tandem, in public schools.[23]

As I will demonstrate in the next chapter, those who favor teaching science rather than myth in public school science classes have so far been able to stave off the imposition of mass preferences on schools by using the federal judiciary to wield the Constitution's First Amendment prohibition of an "establishment of religion." Since public schools are a government institution, and the belief in a supernatural creator is patently a religious doctrine, federal judges, including Supreme Court justices, have issued a string of decisions to ensure that elite rather than mass preferences will determine public school curricula.[24]

Although the preference for teaching science, rather than religion, in public schools would thus seem to be victorious, the victors in the controversy are still uneasy. American intellectuals tend to believe strongly in democracy. Political scientists and philosophers publish dozens of books and articles annually advocating the perfection of democratic institutions; it is virtually impossible to find an author who wishes to do away with the consent of the governed as the legitimating principle of government. But social theorists who believe in democracy and yet endorse the teaching of science, and only science, live with an ideological contradiction. The people clearly desire to have their children taught creationism alongside Darwinism, and have wanted it so for decades. Advocates of secular education cannot endorse democratic government, yet block the consistent and intense preferences of the people, without branding themselves as hypocrites on a rather large scale. Thus, the extensive literature warning that no society should "take the power to define and teach science out of the hands of the scientists" usually contains an embarrassed silence about this policy's inconsistency with democratic theory.[25] Meanwhile, creationists often voice the populist credo that "majorities should rule!"[26]

There is, however, an obvious solution to the problem of the tension between science and democracy in this particular policy area. If public attitudes can be changed so that mass opinion comes more into line with elite opinion, then the ideological contradiction will dissipate. There is precedent for similarly large changes in public sentiment over a relatively brief period of time. Public opinion about gay marriage, for example, has moved from being solidly opposed to cautiously in favor over the past two decades.[27] If such an intense social prejudice can be displaced by tolerance in less than a generation, then it is plausible that attitudes toward evolution can also be modified.

Pro-science social commentators are therefore not just supportive of teaching the theory of natural selection because it will make for a more science-literate, and therefore more employable, citizenry. They also endorse the teaching of the Modern Synthesis as part of a wholesome curriculum because it will, they hope, increase the credibility of the theory itself. Thus, observers who are in favor of biology teaching blame "inadequate training in scientific literacy and an underdeveloped epistemic understanding of science" for the lack of endorsement of "the basic premises of evolution."[28] If that is the problem, then the solution would be to improve the teaching of the theory at both the high school and college levels.[29] And, in fact, various scholars have come up with plans for new ways to teach the subject.[30]

Whether they have novel ideas for a science curriculum or not, most of the pro-science people agree on one position. Schools should NOT, as the proponents of Intelligent Design advocate, "teach the controversy."[31] That is, biology teachers should not be required to discuss the alleged weaknesses in the theory or to lay out the arguments of creationism (Young Earth or Intelligent Design) as an alternative. To do so would only "confuse students and suggest a scientific equality between these different approaches to explaining the biological phenomena of speciation."[32] It would lend legitimacy to "bad math, the lack of respect for evidence, the inculcation of ignorance, and the desire to include non-natural explanations for natural phenomena," which are the "four horsemen who bring death to scientific inquiry."[33] Furthermore, "It has become a matter of social responsibility for biologists, who are the experts in the field, to ensure that correct science is taught to the next generation," but "claims of 'Intelligent Design' are so far from science that we cannot have meaningful discussion of whether claims from ID creationism constitute science *no matter how we demarcate the boundary between science and non-science*."[34] Or, as philosopher Philip Kitcher succinctly states the position, "Lying to the young is not a healthy practice."[35]

It is entirely natural that science-oriented, rather than religion-oriented, commentators would prefer to teach science, and only science, to children. Nevertheless, there is a usually unspoken but problematic assumption underlying their opinions: Students will learn what we try to teach them. The generally unstated assumption is that adult Americans reject evolution because they were taught it badly, or taught outright creationism, in their youths. Conversely, if today's students are taught scientific theory in a competent manner, tomorrow's public opinion surveys will be more agreeable.

The evidence underlying that assumption up to this point in time, however, has not been encouraging. The way to determine if scientific education is effective is to administer before-and-after tests (or, in more technical language, pre-tests and post-tests). That is, survey the knowledge and opinions of students on the first day of the semester, administer the identical survey on the last day of the semester, and compare the two sets of responses. If the class has been effective, there will be evident changes in opinion from pro-creationism to pro–natural selection. In fact, if the teacher has been perfectly effective, there will be no creationist sentiment remaining; the Modern Synthesis will be the unanimous theory of choice.

A few teachers have thus tested the impact of their biology courses, and have been unpleasantly surprised by the results. For example, in 1993 Roger Short gave his medical students at Monash University in Australia before-and-after surveys, in between times presenting evolutionary theory and evidence over a semester. He reported, "To my utter dismay, there were no statistically significant changes in any of the answers to any of the questions."[36] Each fashioning his or her own inquiries, Chinsamy and Planany,[37] Anderson,[38] Lawson and Weser,[39] Sinclair and Pendarvis,[40] and Kottler[41] came to the same general conclusions. Their classes produced some change in beliefs at the margins, but no statistically significant movement toward accepting evolution by those who had initially rejected it. One or two of these results might be attributed to poor teaching, but for such a consistent finding we must turn to another explanation. The conclusion to be drawn from the evidence so far is that the confidence that many pro-science citizens have in the power of education to pry young people away from belief in some form of creationism may be unfounded.

In the previous chapter, I reported on some of the psychological research that describes several inborn tendencies of human cognition, leading to the stubborn resistance with which many people fend off the knowledge of scientific biology. That research, coupled with the evidence from before-and-after tests on the futility of attempting to change religious views with education, would tend to endorse E. Margaret Evans' conclusion from her own research into the beliefs of fundamentalists: "Creationism is here to stay."[42]

But there is reason to think that the pessimism might be premature.

First, a large minority of citizens, not to mention a near unanimity of biologists, have accepted the Modern Synthesis perspective as a true picture of the world. Were they not born intuitive theists? Before 1859, an overwhelming majority of Western citizens were creationists. Today, natural selection is accepted by a significant percentage. Something must have happened to change so many people's minds. If so, then some minds, at least, can be changed. The trick is to discover a means to educate the mistaken majority so that its opinions are brought into line with the views of the accurate minority.

Second, not all pedagogical strategies for imparting the truths of natural selection have been tried. Perhaps there is a better way of teaching evolutionary biology, one that defeats or evades the "natural" tendencies of many people to reject the scientific worldview.

## Research Design

During the spring 2011 semester at the University of Texas at Austin, I taught two undergraduate courses. The first was "Introduction to American and Texas Politics," a lower-division course that every student must take in order to graduate. Because it is required of everyone, its seats are filled by students of all majors and every type of educational background. The second was "The *Origin of Species* and the Politics of Evolution," an upper-division government class—thus ensuring that most of the students would be government majors—that was cross-listed with the university's "great books" program—thus ensuring a sprinkling of students from other majors. Enrollment in each class was slightly shy of 200.

The "Evolution" class was my experimental group. I required that its members fill out a questionnaire on the first and last day of the semester. The questionnaire contained a variety of demographic inquiries, as well as a number of political position questions (about abortion and tax policy, for example). Scattered among these questions were the same four that I had piggybacked on to the Texas poll the year before and that I discussed earlier in this chapter.

The "Introduction" class was my control group. I did not mention evolution during the semester, although no doubt some of the natural science majors encountered it in their other courses. I administered the identical questionnaire on the first and last day of the semester.

This was not, strictly speaking, a scientific control. In a laboratory setting, the students in the two groups would have been matched as to demographics and previous knowledge and beliefs about science. It was, however, impossible to accomplish such matching under the given circumstances. The students in both groups were roughly the same in that they were all young Texans who had been admitted under the same set of criteria to one institution. But because the students in the "Intro" course had arrived more or less randomly (and often unwillingly) in a required class, whereas the members of the "Evolution" class were mostly government majors and self-selected for interest in the topic, there were clear differences between the two samples. Nevertheless, it will be of some value to compare whatever changes are evident in the members of the "Evolution" class with whatever changes show up in the members of the class that was not exposed to the same material.

I thus constructed a study that was as close to a natural experiment as I could manage. My hypothesis was that creationist students in the experimental group would change their opinions to endorse the theory of natural selection and that the students in the control group would either not change at all or change to a lesser extent than those in the experimental group.

My pedagogical strategy, however, differed considerably from the approach taken by other professors who have run more or less the same experiment. They were natural scientists teaching in biology or medical courses, and their approach, as nearly as I can make out from their articles, was to teach the theory of natural selection as settled scientific truth. In contrast, I am a political scientist, and my strategy was to proceed according to the method the partisans of Intelligent Design advocate—to "teach the controversy."

The first third of the semester was largely given over to learning about the philosophy of science, how scientific thinking differs from ordinary, "natural" thinking, to placing Darwin in personal, social, and historical context, and to giving *The*

*Origin of Species* a close reading. I did not simply assign Darwin's book; I devoted one class period (an hour and fifteen minutes) to analyzing his argument and reading important passages aloud, the students following along with their own copies. The second third of the course consisted of a consideration of three politically relevant controversies within evolutionary biology—punctuated equilibrium, sociobiology, and the question of whether, if the "tape of life" were to be rewound and allowed to unspool again, human beings would evolve once again.

In the final third of the course, I presented the controversy between the scientific theory of natural selection and the theory, if it can be called that, of creationism/Intelligent Design. Because my own biases are strongly pro-evolution, and came out clearly in my lectures, I went to some lengths to be evenhanded when presenting this clash of ideas. I assigned a book and two essays by three luminaries of ID (Phillip Johnson,[43] William Dembski,[44] and Michael Behe[45]). I devoted one class session to showing the students a documentary made at the behest of the Discovery Institute, the most important ID organization (and one with which Johnson, Dembski, and Behe are associated). I brought in Robert Koons, a professor in the University of Texas philosophy department who is also associated with the Discovery Institute, to give a guest lecture. (In a polite and scholarly style, Koons told the students that what I had told them was wrong.) Several times I made statements in class to the effect that we (graduate-student teaching assistants graded most final essays) did not care on which side of the controversy the students came down; we only cared about the quality of their essays. On the syllabus, after describing the final essay assignment—evaluate the arguments of both sides of the controversy—I printed the following message:

> Since by this time in the course my own prejudices should be obvious, I want to repeat here my caution that you will not be rewarded for agreeing with me, nor punished for disagreeing. I am interested in the quality of your arguments, not in your conclusions.

The students in this class, then, were required both to learn the details of Darwin's theory and to be exposed to some disagreements within modern evolutionary biology, before being immersed in "the controversy" between ID and the Modern Synthesis. Did that method of teaching about the scientific theory convince more minds than attempting to teach it as factual truth?

Two methodological points are worth attention. First, the students were asked to put their names on both surveys, being vigorously assured that their identities would never be made public. Thus knowing who circled what responses on each survey, I was able to track the development of opinion over the semester. That is, I could not only report the raw numbers and percentages of respondents who chose each answer in each survey, but could identify who changed their minds, and in what direction, for both the experimental and control courses.

Second, the number of respondents who filled out individual questions on the survey varied somewhat from question to question. Some students may have missed the first survey because they registered late for the class. Some may have missed the second because they were ill or otherwise absent that last day. Some may have taken both surveys but for some reason failed to answer a given question on either the first or the second. By the time these various glitches and absences were accounted for, the total number of usable respondents in each class had been reduced by about half. In courses with enrollments of about 200, then, I could accurately track the opinions of roughly 100 from each class. Moreover, the relationship of the raw numbers to the percentages varied slightly from question to question, because the size of the denominator was not the same for each one.

## Results

For help with the analysis of the data in this section, I turned to another of my colleagues. What follows draws on research I conducted with Tse-min Lin in 2013 and 2014 for an article in a professional journal.

On the anonymous teacher evaluations that the students were asked to fill out during the final week of the semester, the class received high marks. All but one of the student respondents pronounced themselves pleased with both my style of instruction and the substance of the knowledge they had absorbed over the semester. The one dissenter told me, in sum, that my soul will burn in hell. But given the incendiary nature of the subject matter, and the generally conservative Christian atmosphere of Texas, I found this ratio of approval to disapproval to be encouraging.

More importantly, the results of the before-and-after tests do show that using a "teach the controversy" pedagogical strategy seems to move opinion toward acceptance of the scientific theory, but the interpretation of the data does not lead to a clear-cut conclusion. Table 4.4 displays for both classes the cross-tabbed relationship of the answers to the first and second iterations of the question "Which of the following statements comes closest to your views on the origin and development of life on Earth?" It is clear that there are only marginal movements between the two surveys administered to the "Introduction" class.

Which of the following statements comes closest to your views on the origin and development of life on earth?

A. Life on earth has existed in its present form since the beginning of time.
B. Life on earth has developed over millions of years, and with no guidance from God.
C. Life on earth has developed over millions of years, but with a guiding hand from God.
D. Don't know.

**TABLE 4.4** Movements in Opinion Concerning the Origin and Development of Life on Earth

"Introduction" Class

|  |  | \multicolumn{5}{c}{Second Survey} |
|---|---|---|---|---|---|---|
|  |  | A | B | C | D | Total |
| First Survey | A | 2 | 0 | 2 | 0 | 4 |
|  | B | 1 | 15 | 1 | 1 | 18 |
|  | C | 1 | 1 | 66 | 1 | 69 |
|  | D | 0 | 3 | 4 | 4 | 11 |
|  | Total | 4 | 19 | 73 | 6 | 102 |

"Evolution" Class

|  |  | \multicolumn{5}{c}{Second Survey} |
|---|---|---|---|---|---|---|
|  |  | A | B | C | D | Total |
| First Survey | A | 0 | 0 | 1 | 1 | 2 |
|  | B | 0 | 37 | 2 | 1 | 40 |
|  | C | 0 | 7 | 32 | 2 | 41 |
|  | D | 1 | 6 | 3 | 2 | 12 |
|  | Total | 1 | 50 | 38 | 6 | 95 |

Specifically, there is a net gain of only one (from 18 to 19) in the scientific category. In comparison, the movements are more apparent for the "Evolution" class. On the first day of the semester, forty students were ready to endorse the scientific answer, "Life on Earth has developed over millions of years, and with no guidance from God." By the end of the semester, an additional thirteen had moved over to that opinion. With a loss of three to other opinions, the net gain in the scientific category is ten (from 40 to 50).

The greater number of scientific believers is attributable to some opinion churn, with most, but not all, of the overall movement toward endorsing the scientific theory. That is, of the forty pro-science students on the first day, thirty-seven were still in that category on the last day, two had moved over to endorsing evolution "with a guiding hand from God," and one professed not to know. Meanwhile, seven of the forty-one who believed in the hand of God on the first day had endorsed secular evolution by the last, and six of the twelve in the "Don't know" category had moved over to endorse the scientific theory by the last day.

Tables 4.5 and 4.6 show the same general pattern. In comparison with the "Introduction" class, the "Evolution" class exhibited a movement away from

**TABLE 4.5** Movements in Opinion Concerning the Evolution of Human Beings from Animals

"Introduction" Class

|  |  | Second Survey |  |  |  |
|---|---|---|---|---|---|
|  |  | A | B | C | Total |
| First Survey | A | 43 | 4 | 3 | 50 |
|  | B | 2 | 31 | 0 | 33 |
|  | C | 9 | 3 | 7 | 19 |
|  | Total | 54 | 38 | 10 | 102 |

"Evolution" Class

|  |  | Second Survey |  |  |  |
|---|---|---|---|---|---|
|  |  | A | B | C | Total |
| First Survey | A | 69 | 1 | 0 | 70 |
|  | B | 4 | 7 | 4 | 15 |
|  | C | 7 | 0 | 2 | 9 |
|  | Total | 80 | 8 | 6 | 94 |

theistic evolution and not knowing toward secular evolution. with net gains in the scientific category amounting to ten (from 70 to 80) and seven (from 45 to 52), respectively.

Please tell us whether you agree or disagree with the following statement (that is, circle the response that best represents your view): "Human beings, as we know them today, developed from earlier species of animals."

A. True
B. False
C. Don't know

Which of the following statements comes closest to your views on the origin and development of human beings?

A. Human beings have developed over millions of years from less advanced forms of life, but God guided the process.
B. Human beings have developed over millions of years from less advanced forms of life, and God had no part in the process.

C. God created human beings pretty much in their present form at one time within the last 10,000 year or so.
D. Don't know

There is thus an unambiguous overall movement, after being exposed to "teach the controversy" pedagogy, from either believing in "theistic evolution" or professing ignorance to endorsing secular evolution. Having identified this pattern, we then turned to the question of whether the pattern is robust enough to establish statistical significance when it is compared with the smaller churn in the "Introduction" class. To our disappointment, the evident movement in the tables was not strong enough to establish statistical significance.[46] Despite the apparently clear trends exhibited in the "Evolution" class compared with the "Introduction" class, the sought effect remains elusive in the statistical sense.

Nevertheless, the general conclusion to be gleaned from an analysis of Tables 4.4, 4.5, and 4.6 would seem to be that this one experiment in using the "teach the controversy" means of conveying the theory of evolution does inspire hope that such an approach can change minds. The data are imperfect and do not

**TABLE 4.6** Movements in Opinion Concerning the Origin and Development of Human Beings

"Introduction" Class

|  |  | Second Survey |  |  |  |  |
|---|---|---|---|---|---|---|
|  |  | A | B | C | D | Total |
| First Survey | A | 30 | 3 | 8 | 0 | 41 |
|  | B | 1 | 17 | 0 | 2 | 20 |
|  | C | 5 | 0 | 21 | 2 | 28 |
|  | D | 4 | 2 | 2 | 2 | 10 |
|  | Total | 40 | 22 | 31 | 6 | 99 |

"Evolution" Class

|  |  | Second Survey |  |  |  |  |
|---|---|---|---|---|---|---|
|  |  | A | B | C | D | Total |
| First Survey | A | 18 | 10 | 2 | 3 | 33 |
|  | B | 3 | 41 | 0 | 1 | 45 |
|  | C | 5 | 0 | 4 | 0 | 9 |
|  | D | 4 | 1 | 0 | 3 | 8 |
|  | Total | 30 | 52 | 6 | 7 | 95 |

permit loud proclamations. More research is necessary, but at least we now have some encouraging avenues of investigation.

## Discussion

The conclusion of this study—and of this chapter—must be to endorse Michael Riess's argument that creationism is not a simple misconception, but part of a "worldview"—what philosophers would call an "ontology"—that encompasses much more than the acceptance of empirical facts. His reminder that "[o]ne rarely changes one's worldview as a result of formal teaching, however well one is taught"[47] is a statement of fact under most circumstances. Yet the result of this one small effort to "teach the controversy" and measure the result might be to introduce some optimism into the discussion. The occurrence of evident, if statistically insignificant, movement in opinion toward acceptance of the theory of natural selection creates the possibility that there might be a way to persuade at least some members of the public to become more accepting of teaching science in public school science classes.

This is one rather frail set of results from two college classrooms. It should not be taken to suggest that I think the emphasis that evolution educators and such organizations as the National Center for Science Education are putting on instruction in elementary and high schools should be abandoned. The traditional emphasis on teaching science as truthful at the lower educational level should be continued. But perhaps the addition of an alternate approach of bringing in critiques and defenses of orthodox biology might be tried on an experimental basis, at least in advanced placement courses. The public support for biological education could not be much more discouraging. Perhaps the numbers could be improved by adopting a different philosophy of teaching science.

It is, of course, a possibility that education at all levels is irrelevant and that ontological beliefs are the result either of parental instruction or personality imperatives.

But let us not assume the worst without further research.

Whatever the final conclusion about the possibility of modifying ontological beliefs, it will bring us back to the problem for democracy. If young minds are to change, it appears that they will often have to be changed against the wishes of their parents. But certainly, adopting "teaching the controversy" as a pedagogical strategy would be more, not less, democratic. It would be a gratifying irony if the public could finally be brought to embrace scientific biology through the use of a strategy recommended by creationists.

## Notes

1 Ferris, Timothy, *The Science of Liberty: Democracy, Reason, and the Laws of Nature* (New York: HarperCollins, 2010), 2.

2 Merton, Robert, *The Sociology of Science: Theoretical and Empirical Investigations* (Chicago: University of Chicago, 1973), 269, 277–278.
3 Prindle, David F., *The Paradox of Democratic Capitalism: Politics and Economics in American Thought* (Baltimore: Johns Hopkins University Press, 2006).
4 de Tocqueville, Alexis, *Democracy in America*, Vol. I (New York: Vintage Books, 1835, 1945), 273, 281.
5 Fodor, Jerry and Massimo Piattelli-Palmarini, *What Darwin Got Wrong* (New York: Farrar, Straus and Giroux, 2011); Nagel, Thomas, *Mind and Cosmos: Why the Materialist Neo-Darwinian Conception of Nature Is Almost Certainly False* (Oxford: Oxford University Press, 2012).
6 Bishop, George F., Randall K. Thomas, Jason A. Wood, and Misook Gwon, "Americans' Scientific Knowledge and Beliefs about Human Evolution in the Year of Darwin," *Reports of the National Center for Science Education*, 30, no. 3 (May-June 2010), 16–18; Bishop, George F., Randall K. Thomas, and Jason A. Wood, "Measurement Error, Anomalies, and Complexities in Americans' Beliefs about Human Evolution," *Survey Practice*, (October 2010).
7 Kellstedt, Lyman A. and Corwin E. Smidt, "Doctrinal Beliefs and Political Behavior: Views of the Bible," in David C. Leege and Lyman Kellstedt, eds., *Rediscovering the Religious Factor in American Politics* (Armonk, N.Y.: M.E. Sharpe, 1993), 177.
8 Blanchard, Troy C., John P. Bartkowski, Todd L. Matthews, and Kent R. Kerley, "Faith, Morality and Mortality: The Ecological Impact of Religion on Population Health," *Social Forces*, 86, no. 4 (June 2008), 1591–1620; Mazur, Allan, "Believers and Disbelievers in Evolution," *Politics and the Life Sciences*, 23, no. 2 (November 2005), 55–61; Steensland, Brian, Jerry Z. Park, Mark D. Regnerus, Lynn D. Robinson, W. Bradford Wilcox, and Robert D. Woodberry, "The Measure of American Religion: Toward Improving the State of the Art," *Social Forces*, 79, no. 1 (September 2000), 291–318; Ellison, Christopher G. and Marc A. Musick, "Southern Intolerance: A Fundamentalist Effect?" *Social Forces*, 72, no. 2 (December 1993), 379–309.
9 McDaniel, Eric L. and Christopher G. Ellison, "God's Party? Race, Religion, and Partisanship over Time," *Political Research Quarterly*, 61, no. 2 (June 2008), 180–191; Hempel, Lynn M. and John P. Bartkowski, "Scripture, Sin and Salvation: Theological Conservatism Reconsidered," *Social Forces*, 86, no. 4 (June 2008), 1647–1674; Evans, E. Margaret, "Cognitive and Contextual Factors in the Emergence of Diverse Belief Systems: Creation versus Evolution," *Cognitive Psychology*, 42 (2001), 217–266.
10 Ecklund, Elaine Howard, Jerry Z. Park, and Phil Todd Veliz, "Secularization and Religious Change among Elite Scientists," *Social Forces*, 86, no. 4 (2008), 1805–1839.
11 Ellison, Christopher G. and Kathleen A. Nybroten, "Conservative Protestantism and Opposition to State-Sponsored Lotteries: Evidence from the 1997 Texas Poll," *Social Science Quarterly*, 80, no. 2 (June 1999), 356–369; Darren E. Sherkat and Christopher G. Ellison, "The Cognitive Structure of a Moral Crusade: Conservative Protestantism and Opposition to Pornography," *Social Forces*, 75, no. 3 (March 1997), 957–982.
12 Leege, David C. and Lyman Kellstedt, "Religious Worldviews and Political Philosophies: Capturing Theory in the Grand Manner through Empirical Data," in Leege and Kellstedt, *Rediscovering the Religious Factor*, op. cit., 216–231.
13 Gallup website, June 1, 2012. http://www.gallup.com/poll/155003/ hold-creationist-view-human-origins.aspx: accessed June 20, 2014.
14 Mazur, Allan, "Believers and Disbelievers in Evolution," *Politics and the Life Sciences*, 23, no. 2 (2005), 55–61.

15 Bishop et al., "Americans' Scientific Knowledge," op. cit., and "Measurement Error," op. cit.
16 Putnam, Robert D. and David E. Campbell, *American Grace: How Religion Divides and Unites Us* (New York: Simon and Schuster, 2010), 27, 272.
17 Prindle, David and Brian Roberts. 2010. "Was Lewis Black Right? Public Opinion about Evolution in Texas." American Political Science Association Annual Meeting Paper. Available at SSRN: http://ssrn.com/abstract=1642448: accessed June 24, 2014.
18 Wilson, G.D., "A Dynamic Theory of Conservatism," in G.D. Wilson, ed., *The Psychology of Conservatism* (London: Academic Press, 1973), 259, quoted in Jost, John T., Arie W. Kruglanski, Jack Glaser, and Frank J. Sulloway, "Political Conservatism as Motivated Social Cognition," *Psychological Bulletin*, 129, no. 3 (2003), 339–375 (Wilson quoted on p. 347).
19 Jost et al., ibid., 345.
20 Ibid., 347.
21 Ibid., 352.
22 Putnam and Campbell, *American Grace*, op. cit., 547, 540.
23 Berkman, Michael and Eric Plutzer, *Evolution, Creationism, and the Battle to Control America's Classrooms* (New York: Cambridge University Press, 2010), 35–39.
24 *Epperson v. Arkansas*, 393 U.S. 97 (1987); *Edwards v. Aguillard*, 482 U.S. 578 (1987); *Kitzmiller v. Dover Area School District* (2005), available at www.pamd.uscourts.gov/kitzmiller.342.pdf.
25 Davis, Michael J., "Religion, Democracy, and the Public Schools," *Journal of Law and Religion*, 25 (2009–2010), 38; for a partial exception to the generalization I make here, see Berkman and Plutzer, *Evolution*, op. cit., 8–13.
26 Humes, Edward, *Monkey Girl: Evolution, Education, Religion, and Battle for America's Soul* (New York: HarperCollins, 2007), xi, xii, 207.
27 Gallup website, reporting on findings of a recent survey, May 8, 2012.
28 Hofer, Barbara K., Chak Fu Lam, and Alex DeLisi, "Understanding Evolutionary Theory: The Role of Epistemological Development and Beliefs," in Roger S. Taylor and Michel Ferrari, *Epistemology and Science Education: Understanding the Evolution vs. Intelligent Design Controversy* (New York: Routledge, 2011), 95–96.
29 Ibid., 105.
30 Wilensky, Uri and Michael Novak, "Teaching and Learning Evolution as an Emergent Process: The BEAGLE Project," in Taylor and Ferrari, *Epistemology and Science Education*, op. cit., 213–242; Alters, Brian, "Evolution in the Classroom," in Eugenie C. Scott and Glenn Branch, eds., *Not in Our Classrooms: Why Intelligent Design Is Wrong for Our Schools* (Boston: Beacon Press, 2006), 105–129.
31 Forrest, Barbara and Paul R. Gross, *Creationism's Trojan Horse: The Wedge of Intelligent Design* (Oxford: Oxford University Press, 2004), 206.
32 Ferrari, Michel and Roger S. Taylor, "Teach the Demarcation: Suggestions for Science Education," in Taylor and Ferrari, *Epistemology and Science Education*, op. cit., 282.
33 Gilbert, Scott F., "The Aerodynamics of Flying Carpets: Why Biologists Are Loath to 'Teach the Controversy,'" in Nathaniel C. Comfort, ed., *The Panda's Black Box: Opening Up the Intelligent Design Controversy* (Baltimore: Johns Hopkins University Press, 2007), 50.
34 Sarkar, Sahotra, *Doubting Darwin? Creationist Designs on Evolution* (Malden, Mass.: Blackwell Publishing, 2007), 156, 162.
35 Kitcher, Philip, *Science in a Democratic Society* (Amherst, N.Y.: Prometheus Books, 2011), 228.

36 Short, R.V., "Darwin, Have I Failed You?" *The Lancet*, 343 (February 26, 1994), 8896, ProQuest p. 528.
37 Chinsamy, Anusuya and Eva Planany, "Accepting Evolution," *Evolution*, 62, no. 1 (2007), 248–254.
38 Anderson, Mike L., "The Effect of Evolutionary Teaching on Students' Views of God as Creator," *Journal of Theology for Southern Africa*, 87 (1994), 69–73.
39 Lawson, Anton E. and John Weser, "The Rejection of Nonscientific Beliefs about Life: Effects of Instruction and Reasoning Skills," *Journal of Research in Science Teaching*, 27 (1990), 589–606.
40 Sinclair, Anne and Murray Pendarvis, "The Relationship between College Zoology Students' Beliefs about Evolutionary Theory and Religion," *Journal of Research and Development in Education*, 30, no. 2 (1997), 118–125.
41 Malcolm Kottler's study reported in Kitcher, *Science in a Democratic Society*, op. cit., 26.
42 Evans, E. Margaret, "Beyond Scopes: Why Creationism Is Here to Stay," in Karl S. Rosengren, Carl N. Johnson, and Paul L. Harris, eds., *Imagining the Impossible* (Cambridge: Cambridge University Press, 2000), 305–333.
43 Johnson, Phillip E., *Darwin on Trial*, 2nd ed. (Downers Grove, Ill.: InterVarsity Press, 1993).
44 Dembski, William A., "Intelligent Design as a Theory of Information," in Robert F. Pennock, ed., *Intelligent Design Creationism and Its Critics: Philosophical, Theological, and Scientific Perspectives* (Cambridge, Mass.: MIT Press, 2001), 553–574.
45 Behe, Michael J., "Moleculare Machines: Experimental Support for the Design Inference," in ibid., 241–256.
46 To test for statistical significance, we applied the "difference-in-differences" technique to our data. (See Greene, William H., *Econometric Analysis* [Boston: Prentice Hall, 2012], 155–158.) The technique allows us to compare the movements toward accepting scientific theory between the two classes in a statistically rigorous fashion. We conducted the analysis for all three questions. The results are shown in Table 4.7.

**TABLE 4.7** Difference-in-Differences Analysis

|  | Linear Probability Models |||
| --- | --- | --- | --- |
|  | Origin and Development of Life on Earth | Evolution of Human Beings from Animals | Origin and Development of Human Beings |
| Constant | 0.1765*** | 0.4902*** | 0.1961*** |
|  | (0.0441) | (0.0456) | (0.0451) |
| s (Experimental vs. Control Group) | 0.2359*** | 0.2315*** | 0.2678*** |
|  | (0.0631) | (0.0654) | (0.0645) |
| t (Second vs. First Survey) | 0.0098 | 0.0392 | 0.0196 |
|  | (0.0623) | (0.0646) | (0.0637) |
| s·t (Interaction of X and T) | 0.0933 | 0.0742 | 0.0526 |
|  | (0.0892) | (0.0925) | (0.0913) |
| N | 398 | 398 | 398 |
| Adj. $R^2$ | 0.0909 | 0.0790 | 0.0912 |

*Note*: Standard errors in parenthesis. Statistical significance: *** $p < .001$, 2-tailed tests.

The "difference-in-differences" test is based on the linear regression model:

$$Y_{DT} = \beta_0 + \beta_1 D + \beta_2 T + \beta_3 D \cdot T + \varepsilon \quad \varepsilon \sim N(0, \sigma_\varepsilon^2)$$

The dependent variable, $Y_{DT}$, is a dummy variable indicating whether a student chose the scientific answer in response to each of the three survey questions, where $D = 1$ if the student was in the "Evolution" class and $D = 0$ otherwise, and $T = 1$ if the response was given at the second survey and $T = 0$ otherwise. The dummy variables $D$, $T$, and their product $D \cdot T$ also serve as independent variables. Based on this model, the expected values of $Y$ (which in this case amounts to proportions) for the four surveys can be expressed in terms of the regression coefficients:

$$\pi_{00} = E(Y_{00}) = E(Y_{st} \mid D = 0, T = 0) = \beta_0$$
$$\pi_{10} = E(Y_{10}) = E(Y_{st} \mid D = 1, T = 0) = \beta_0 + \beta_1$$
$$\pi_{01} = E(Y_{01}) = E(Y_{st} \mid D = 0, T = 1) = \beta_0 + \beta_2$$
$$\pi_{11} = E(Y_{11}) = E(Y_{st} \mid D = 1, T = 1) = \beta_0 + \beta_1 + \beta_2 + \beta_3$$

where $\pi_{00}$, $\pi_{10}$, $\pi_{01}$, and $\pi_{11}$ are the proportions of respondents choosing the scientific answer in the first survey for the Introduction class, the first survey for the Evolution class, the second survey for the Introduction class, and the second survey for the Evolution class, respectively. Consequently,

$$(\pi_{11} - \pi_{10}) - (\pi_{01} - \pi_{00}) = ((\beta_0 + \beta_1 + \beta_2 + \beta_3) - (\beta_0 + \beta_1)) -$$
$$((\beta_0 + \beta_2) - (\beta_0)) = \beta_3$$

That is, $\beta_3$ measures the difference between $\pi_{11} - \pi_{10}$ (i.e., net gain in proportion of respondents choosing the scientific answer for the Evolution class) and $\pi_{01} - \pi_{00}$ (i.e., net gain in proportion of respondents choosing the scientific answer for the Introduction class). The significance test of $\beta_3$ therefore provides the "difference-in-differences" test of the effect of my "teaching the controversy" experiment.

As shown in Table 4.7, none of the estimates of $\beta_3$ is statistically different from zero.

47 Reiss, Michael J., "The Relationship between Evolutionary Biology and Religion," *Evolution*, 63, no. 7 (2009), 1940.

# 5

# THE JURISPRUDENCE OF EVOLUTION

Ideally, law is philosophy codified. When a legal system—and, by extension, court decisions—are functioning well, a consistent set of general principles is deduced from an even larger set of basic values, and then applied to specific cases. As with other human institutions, law is seldom ideal, because it is subject to the deforming pull of economic interest, ethnic and sexual prejudices, the simple tendency to make mistakes, and the inability to foresee the future. But many people within the legal system are constantly struggling to bring the threatened chaos of legal reasoning into conformity with their own version of ideological consistency. They often succeed pretty well, and so it is possible to discuss the development of doctrines about certain areas of legal problems without doing too much violence to the way courts and judges actually function.

In the American context, the ideal that is constantly before the eyes of serious thinkers is the achievement of a style of Constitutional interpretation that will apply the words of the fundamental law, written during the eighteenth century, to the needs of a twenty-first century society. From the nation's founding moment to the present, however, the construction of an ideal to which all Constitutional thought should aspire has been frustrated by the fact that the American ideology has been constructed of an only rarely acknowledged contradiction.

On the one hand, the ideal of government has been a democratic one, bottomed on the "consent of the people" as the "pure, original fountain of all legitimate government," to quote an oft-repeated metaphor from *The Federalist Papers*.[1] On the other hand, the authors of the Constitution were intensely suspicious of members of the public at large. Humans in general, they believed, were "ambitious, vindictive, and rapacious," and therefore, not to be trusted with power.[2]

Elites were not a reliable guarantor of republican liberties. But even more, ordinary, non-elite citizens were, in Hamilton's words, "subject to the arts of designing men" who would often induce them to support "dangerous innovations in the government and serious oppressions of the minor party in the community."[3] By the last phrase, Hamilton meant property owners, who seemed to be vulnerable to the envy of the mob, and who thus needed a governmental safeguard against democratic excesses. Hamilton's preferred solution to the problem, as he famously explained in *Federalist* #8, was an independent judiciary (that is, one in which the judges would enjoy lifetime appointment), as a check on the dangerously democratic tendencies of Congress.

Hamilton got his independent judiciary, but its institutionalization did not reconcile the contradictory impulses that have governed American political and legal thought ever since. Do we trust the people or not? Do we allow the people to choose the republic's policies or not? Under what circumstances, and applying what principles, should the public have its way, even if judges, scholars, journalists, and an intense minority believe the public's ideas to be destructive of prosperity, safety, or liberty? Thus, scholars in this area often debate the problem of the "counter-majoritarian difficulty," which centers on whether the justification for judicial review by a non-elected judiciary is persuasive.[4] When they do conclude that judicial review is justified, they typically argue that it is another way of implementing democracy, complementary to, rather than contradicting, the implementation of the people's will through the legislature.[5]

The jurisprudence of evolution, however, is more like Hamilton's original discussion of an independent judiciary—oriented to the question of how to protect one segment of society from democratic meddling. As I demonstrated in Chapter Four, the public consistently, and by large majorities, favors the teaching either of creationism alone or of creationism alongside scientific biology, in the public schools. Yet the position of educated elites in general and scientists in particular, as I discussed in Chapter Three, is that public education must never be organized so as to acquiesce to those democratic desires. The problem for judges, as representatives of those educated elites, is to contrive a "jurisprudential regime" that will justify persistently anti-democratic rulings within the rhetoric of the democratic Constitution.[6]

In virtually every textbook on American government, in the chapter on the judicial system, there is a section on styles of Constitutional interpretation. The model that is offered to undergraduates in these texts is almost invariably dichotomous, describing two types of judges, those who are "strict constructionists" and those who are "activist judges." According to the texts, strict constructionists content themselves with applying the literal meaning of the Constitution's words, while activists try to create new interpretations to further their social and economic agendas.

As scholars of jurisprudence, as opposed to the authors of American Politics textbooks know, this understanding of the way judges go about applying the Constitution is a misleading distortion. Historically, there have not been two ways of approaching the Constitution; there have been many. Every advocate of this or that judging doctrine is trying to apply a given theory of Constitutional interpretation. Over the last several decades, the phrase "activist judge" has become an epithet, and most nominees to the Supreme Court attempt to position themselves as strict constructionists in their Senate hearings. But no judge is a strict constructionist, although some believe they are. Because there is no interpretation of the Constitution that is written in the stars, unarguably manifest for all to see, every justice must act within one or another theory of interpretation. In that sense, every justice is an activist. The task for legal scholars and concerned citizens should not be to decide who is or is not an activist, but to try to discern which of the many theories of Constitutional jurisprudence underlies any given decision or series of decisions.

Ever since New York's Chancellor Kent published the first volume of *Commentaries on American Law* in 1826, American legal thinkers have been attempting to come up with a consistent, systematic theory of Constitutional interpretation that would ground the legitimacy of government on the consent of the governed and yet forbid the governed to take actions that the theorist would regard as reprehensible.[7] (Kent's purpose, like Hamilton's, was to prevent the people from abrogating the rights of property.) Although some interpretive doctrines have dominated in some eras of our history, none has ever reigned unchallenged. In treatises, law review articles, and court decisions, scholars and judges have fought a continuing battle of interpretation that has accompanied the political battle going on in other areas of American society.

One important contemporary doctrine of interpretation, for example, is "original understanding," identified with the late federal judge Robert Bork and, today, with Constitutional scholar Lino Graglia. According to Bork, justices should interpret the Constitution in the manner that its words would have been understood in 1787 (or, in the case of the Fourteenth Amendment, 1868). This doctrine would, if followed, be "politically neutral" and would thus spare the Court from the public suspicion that it was simply imposing its preferences on other people.[8] If the justices had always followed this rule, he argues, the Supreme Court would not have made the terrible mistakes of the decisions in *Dred Scott v. Sandford* (1857), *Lochner v. New York* (1905), and *Roe v. Wade* (1973).[9]

A contrary doctrine of Constitutional interpretation, identified with a number of legal theorists, might be termed "pursue democracy." Implementation of the public's preference is, according to this theory, the only standard against which statutes are to be judged. As Justice Stephen Breyer wrote in a 2005 book, "Legislation in a delegated democracy is meant to embody the people's will, either directly ... or indirectly.... Either way, an interpretation of a statute that tends to implement the legislator's will helps to implement the public's will and is therefore consistent with the Constitution's democratic purpose."[10]

These two doctrines of Constitutional interpretation, like the others that have engendered larger or fewer numbers of disciples among legal scholars, wax and wane in influence depending upon the power of their exposition, the preferences of the president who appoints them, and the social circumstances at any given time.[11]

In addition to these large, overall approaches to Constitutional interpretation, however, there have been, at various times in American history, theories about how the document should be applied to more restricted areas. Again, the limited doctrines generally try to reconcile the nation's bedrock democratic commitments with the alarming tendency of ordinary people to endorse tyrannical policies. (Which of the people's opinions support tyrannical government actions varies, of course, with the political values of the theorist.)

The most infamous of these limited approaches to Constitutional interpretation was the doctrine of "substantive due process," which dominated Supreme Court decision making about economic regulation from the 1890s to the 1930s. A beautiful example of the way the alleged implications of Darwinian theory could be applied in real-world situations, this doctrine strained Herbert Spencer's version of Darwin, discussed in Chapter One, through the "due process of law" clause of the Fourteenth Amendment and concluded that regulation of economic activity by legislatures violated moral natural law. Because the justices also believed that the Constitution embodied moral natural law, they felt justified in overturning such state and federal legislative actions as statutes regulating working hours, ensuring that companies paid their employees fairly, and forbidding companies to engage in monopolistic practices.[12]

At the intersection of law, religion, and public policy, also various doctrines jockey for acceptance in both courtrooms and public discourse. There is a general reluctance, born of long and painful experience with religious conflict, to allow government to interfere with the practice of people's faith. There is also a general determination to keep government as separate from religion as possible. But these general desires keep confronting specific instances in which any choice of public policy would seem to violate one or both intentions, as well as contradict the desire to implement the people's will. Thus, the jurisprudence of religion rarely satisfies anyone. It is particularly unsatisfactory when it deals with the subject of public education.

## First Amendment Blues

In the early summer of 1962, a few months into my fourteenth year, my father moved me, my brother, my two sisters, and our mother from our home in Hermosa

Beach, California, south of Los Angeles, to Murrysville, in western Pennsylvania. My father was by temperament and profession a salesman, and his success in persuading elements of the military-industrial complex to buy the internal components of fighter planes had persuaded his corporate bosses to promote him to regional sales manager of the Northeastern states. While glorying in the new, lushly green semi-rural environment that was so different from the dry, sandy, suburban landscape of my previous California experience, I waited to enter the ninth grade in my new school.

When it came in early September, my first day in the Franklin Regional school contained several surprises. First, entering the ninth grade would have put me into high school back in California, but high school did not begin until the tenth grade in Pennsylvania, so I was still in what was then known as "junior high," and would now be known as "middle school." It seemed almost a demotion. Second, I was caught unprepared when, in the first minutes of the first class of the first Monday, the male teacher told us all to rise and led us in a chorale recitation of the Lord's Prayer. There had been no prayers in the public schools I had attended in California, let alone any that had been directed by teachers. Third, as I recited the words of the Prayer as I had always heard them repeated back in the United Church of Christ congregation my family attended in Manhattan Beach, I stumbled as I began to intone, "and forgive us our debts, as we forgive our debtors." Everyone else was reciting different words. I stopped and listened, and heard those around me say, "and forgive us our trespasses, as we forgive those who have trespassed against us." Much later, I learned that this was the version of the Prayer spoken by Catholics and United Methodists. So Christians could not even agree on the proper wording of their most basic act of devotion.

At any rate, I adjusted easily enough to the new routine. Nevertheless, our collective homesickness for sun, beach, family, and friends, rather than offense that his Protestant children were being made to recite Catholic prayers in school, was undoubtedly my father's motive for quitting his job less than a year later and moving us all back to Hermosa Beach. He took another sales job and made a good living.

However, If I had been paying closer attention to current events during that summer, instead of catching cottontails in Havahart traps in the woods and pulling giant bluegills out of Hays' Pond down the road, I would have had an even bigger surprise. For in June of 1962, three months before I entered the ninth grade, the United States Supreme Court had, in the case of *Engel v. Vitale*, decreed that a state-sponsored prayer in a public school was a violation of the "establishment" clause of the First Amendment, and therefore forbidden.[13] No one in the Murrysville school showed any sign of having heard about the decision. The administration and the teachers continued their devotional routine as if the United States Supreme Court did not exist.

It is possible that they are still doing so. Various researchers have concluded that there were still some school districts in disparate parts of the country continuing to begin morning classes with Christian prayers, decades after the *Engel* decision.[14]

Whether or not parents and school officials abide by the prayer decision and the secularizing decisions that followed, however, many of them continue to resent the Court's modernizing strategy. As the evolution decisions are part of that strategy, it will be useful to summarize its general elements before honing in on the specific topic of evolution, creationism, and First Amendment jurisprudence.

The First Amendment to the Constitution contains two clauses addressing these issues. "Congress shall make no law respecting an establishment of religion" is the first, "or prohibiting the free exercise thereof" is the second. The clauses are conceptual and, often in practice, distinct. But there are cases in which the line between them blurs, and sometimes it has been in the interest of legal counsel to blur them. It is also sometimes difficult to distinguish the second clause from the additional guarantee in the First Amendment of freedom of speech and press. Furthermore, various parties to controversies have brought in an alleged principle, nowhere stated in the Constitution, of a parental right to oversee the raising of their children, including the right to supervise their education. Justices have sometimes accepted this principle as embodied in the founding document. Finally, although the First Amendment applied only to the federal government, since the Supreme Court began to "incorporate" the Bill of Rights into the Fourteenth Amendment in 1925, most of the doctrines that have applied to the federal government pertain equally to the states.

The Supreme Court, over the decades, has jerkily and sometimes inconsistently fashioned a set of doctrines that take the various principles into account, so that there are, by now, some relatively clear guidelines that apply everywhere in the country. Law school professors continue to criticize various Court decisions, and some large interest groups in society, particularly the anti-modernist religious conservatives discussed in Chapters Three and Six, continue to feel that their rights are abused by the justices. Nevertheless, there is an understandable jurisprudence of government and religion that applies directly to the teaching of evolution in the public schools.

The dominant discourse of the "free exercise" clause was articulated eleven years after the Fourteenth Amendment was adopted in 1868. The case (*Reynolds v. United States*) dealt with the validity of the federal law forbidding polygamy in the territories. (Utah and Idaho were not yet states in 1879.) Could the authority of the people's representatives to write marriage laws be abrogated by the religious belief held by the members of the Church of Jesus Christ of Latter-day Saints (the Mormons) that marriage between a man and several women was divinely sanctioned?

In its decision, the Court quoted Thomas Jefferson to the effect that "the legislative powers of the government reach actions only, and not opinions." The Mormons could, held the justices, go on believing in polygamy as much as they wanted, but government had the legitimate authority to forbid them to put it into practice. As long as the law was neutral on its face—that is, as long as it applied to everyone, and not just Mormons—government could make rules that nevertheless affected some religious organizations and beliefs more than others.[15]

In the twentieth century, the Court wound its way, not always in a straightforward manner, to two principles that fleshed out the basic insight of *Reynolds*. First, under the "no harm" principle, religious entities, as much as other citizens, could be prevented from injuring others, and were thus subject to all laws of general applicability. This principle did not permit governments to impose a religious or patriotic ritual on a non-believer—it could not, for example, require students to participate in the Pledge of Allegiance in school.[16] Under the no-harm principle, however, religious individuals could be forced to obey, for example, the anti-drug laws.[17] The reasoning behind the principle was stated lucidly by Justice O'Connor in a 1988 majority opinion in the *Lyng* case:

> [G]overnment simply could not operate if it were required to satisfy every citizen's religious needs and desires. A broad range of government activities—from social-welfare programs to foreign aid to conservation policies—will always be considered essential to the spiritual well-being of some citizens, often on the basis of sincerely-held religious belief.... The First Amendment must apply to all citizens alike, and it can give to none of them a veto over public programs that do not prohibit the free exercise of religion.[18]

Second, under the "non-persecution" principle, governments could not single out specific religious practices or beliefs for suppression or punishment. In 1993, for example, the Court tossed out an ordinance passed by the city of Haileah, Florida outlawing the "sacrifice of animals." Since the statute did not prevent meat-packing plants from engaging in the slaughter of animals for human consumption, it had plainly been passed to suppress the religious rituals of a group of Santerians. Thus, singling out a religious practice that would not have been proscribed if it had been performed within a secular context, the act was discriminatory and thus violated the principle.[19]

The two principles, however, can be confused and rendered non-operative when justices decide to apply the third, unwritten principle of parental rights. In *Wisconsin v. Yoder* in 1972, the Court majority ruled that the Amish and Mennonites could ignore, for religious reasons, a state law requiring children to attend high school until they were sixteen. The law was intended to preserve children from harm, because in the modern world a lack of education is likely to result in poor economic prospects. It was non-discriminatory, because it applied to all residents of the state equally. But the two religious groups argued that the modern secular high school was contrary to their religion and way of life and that forcing their children to attend such an institution might "endanger their own salvation and that of their children."[20] The Supreme Court agreed, ruling that the state had failed to show that it had a "compelling interest" in shepherding its young citizens through an education beyond the eighth grade and that therefore the interests of the parents in preserving their religiously motivated atavistic lifestyle must be respected.[21]

The *Yoder* decision has attracted a good deal of criticism from legal scholars, and its "hybrid rights" approach has been rejected as unworkable by the lower courts.[22] In 1987, in the *Mozert* case, a three-judge panel of the federal courts refused to be persuaded by the argument that, under the *Yoder* precedent, fundamentalist Christian parents should be allowed to decide which parts of the curriculum of public schools would be presented to their children, and which parts the children could avoid.[23]

Taken together, these two cases are useful in illustrating the sorts of problems that the various principles invoked in the free-exercise debate do not solve. If religious people believe that their creation myths are literally true—and this goes for the dozens of others from various faiths around the world as much as it goes for fundamentalist Christians—then to subject their children to a science curriculum that, by implication, defines that myth as fictional is to compromise their ability to pass their beliefs and values on to their children. A "free exercise" in which parents cannot instruct children is free in only an ironic sense.

Conversely, if evolution is a doctrine the knowledge of which will save children from the harm of ignorance, and if it is neutrally applicable to all children, then religious objection should grant no exemption from general laws.

Further, if some parents have the right to protect their children from the baleful effects of modernist knowledge, then all should have the same right. If the Amish have the right to withdraw their children from the modernizing institution of the public school, then it is hard to see why fundamentalist Christians do not have the right to shield their children from the modernizing ideology of the theory of natural selection. Thus, although the legal doctrines associated with free exercise are clear, their applications to specific cases have created a serious confusion. Court decisions on the free-exercise clause since the 1980s, and congressional participation in the process of determining which religious claims fall under the clause, have not reduced the ambiguity.

But the "free exercise" clause is only part of the problem.

The jurisprudence deriving from the "establishment of religion" clause goes back to a letter written by Thomas Jefferson to the Danbury Baptist Association of Connecticut in 1802:

I contemplate with solemn reverence that act of the whole American people which declared that their legislature should "make no law respecting an establishment of religion, or prohibiting the free exercise thereof," thus building a wall of separation between Church and State.[24]

The "wall" in the passage is, of course, a metaphor, and metaphors, being rhetorical, make frail building blocks for legal doctrines. The history of the Supreme Court's efforts to apply the metaphor to actual cases is one that has kept legal scholars busy writing critical appraisals. Over decades, the justices' efforts to make clear how the wall does or does not apply to public financing for parochial schools,[25] the display of the Ten Commandments in county courthouses,[26] state tax exemptions for religious periodicals,[27] and similar questions produced a set of malleable rules for deciding what was permitted. In 1971, the justices fashioned the "Lemon Test" for determining acceptability under the metaphor:

a. *The government's action* **must have a secular legislative purpose.**
b. *The government's action* **must not have the primary effect of either advancing or inhibiting religion.**
c. *The government's action* **must not result in an "excessive government entanglement" with religion.**[28]

But the rules for deciding whether a government action was acceptable under the Lemon Test, like the rules for deciding what was pornography, proved to be awkward and unsatisfactory when judges tried to apply them. The Supreme Court continued to tinker with, and sometimes ignore, *Lemon*, as its composition changed over the decades. In a book published in 2011, two Constitutional scholars identified seven formulas that had been applied to establishment cases since *Lemon*:

1. Separationism
2. Accommodationism
3. Neutrality
4. Endorsement
5. Coercion
6. Equal treatment
7. History or tradition[29]

So the holding of any given combination of justices on any given case is impossible to predict with accuracy. Nevertheless, as an overall direction, the Court manifestly leans toward insisting that government in general, and schools in particular, be secular institutions. Thus, if any governmental body, especially a public school, is coercively imposing an official religious ceremony or doctrine on citizens who have not requested it, then the action is likely to be held unconstitutional. Students can pray all they want as individuals, but a school cannot allow them to pray over a public address system or to a captive audience. Students and teachers can believe that the first book of *Genesis* is an actual history of the cosmos, but they may not include that belief as part of the official instruction of a public school, especially in a science class. A public school must be secular in its curriculum, regardless of the content of the beliefs of most of its teachers and students, and regardless of the intensity with which they hold those beliefs.

While the general secularizing trend is evident in First Amendment jurisprudence, however, the devil, as usual, is in the details. The specific court cases dealing with evolution fall under the doctrines of the free-exercise and establishment clauses, but those doctrines have supplied a sometimes wobbly set of rationales for the decisions. The difficulty of applying ambiguous rules to specific cases is worsened by the difficulty of defining such important auxiliary terms as "science." Nevertheless, judges have made their decisions, and they form a more or less coherent set of rules about the teaching of evolution in the classroom. Those decisions are the motive forces for the political struggle over the public school curriculum that has been ongoing since the 1960s.

## Evolution and the Courts

In 1928, in the backwash of the Scopes trial, the Arkansas legislature passed a law forbidding teachers in public schools "to teach the theory or doctrine that mankind ascended or descended from a lower order of animals."[30] As I discussed in the previous chapter, this law addressed the issue that is most sensitive to religious believers—that the human species specifically, rather than life in general, is the result of natural processes. It is also noteworthy that the law was passed, not by the state's legislators, but by its citizens under the initiative process. It was thus a genuine expression of majority opinion in that state.

In 1966 Susan Epperson, a biology teacher at Little Rock Central High School, with the backing of the American Civil Liberties Union, the National Education Association, the Little Rock Ministerial Association, and at least one parent, filed suit in state court, asking that the law be enjoined from enforcement. She won the case at the lower level, but the Arkansas state supreme court overturned the holding. She then appealed to the federal courts.[31]

Given the U.S. Supreme Court's secularizing preconceptions, the decision was easy. Justice Fortas, writing for the majority, held that "Arkansas' law selects from the body of knowledge a particular segment which it proscribes for the sole reason that it is deemed to conflict with a particular religious doctrine." Citing newspaper advertisements and letters supporting adoption of the statute during the 1920s, Fortas opined that it is "clear that the fundamentalist sectarian conviction was and is the law's reason for existence." Because the public schools were a state institution, the law therefore violated the "no establishment of religion" clause of the First Amendment, as applied to the states by the Fourteenth Amendment.[32]

Justice Black concurred, but for different reasons. The state law, he wrote, was vague, and therefore unconstitutional. But it was a troubling precedent, he thought, for the Court to engage in the practice of telling states what they must put into, or omit, from their public school curricula:

> It would be difficult to make a First Amendment case out of a state law eliminating the subject of higher mathematics, or astronomy, or biology from its curriculum. . . . [T]here is no reason I can imagine why a State is without power to withdraw from its curriculum any subject deemed too emotional and controversial for its public schools.[33]

Both Fortas' majority opinion and Black's concurrence sowed the seeds of future troubles. First, Black understood that the Court was telling the states not only that they were forbidden to impose the religious orthodoxies of citizen majorities on students who might not want to be so burdened, but that state institutions *must* teach subject matter that democratic majorities did not want taught. He was arguing, in other words, that the decision was not defensive, protecting minority viewpoints from the tyranny of the majority, but aggressive, mandating that state governments could not evade, but must adopt, a policy anathema to the majority. Such an aggressive decision was bound to feed the paranoia of large numbers of people, especially in the South, that a modernist elite was taking away their right to govern themselves.

Second, Fortas had written that "the state may not adopt programs or practices in its public schools or colleges which 'aid or oppose' any religion." That sentence was soon interpreted to mean that the problem with the Arkansas law was that it was one-sidedly an imposition of faith-based belief.[34] Arkansas had established a religious doctrine as the one and only one acceptable in a state institution. But suppose—and one can imagine the reasoning of minds all over the South—what was taught in the schools was a *scientific* theory that just happened to be consistent with *Genesis* rather than *The Origin of Species*? Surely, that would pass Constitutional muster. Or, even better, why not mandate the teaching of both sciences, creationist and evolutionary? Explain both theories to the students, and let them make up their own minds. What could be more evenhanded? What could be more Constitutional?

As it happened, there was already a movement among the more sophisticated fundamentalists to create a theory of "creation science" that would position *Genesis* as parallel to, rather than in opposition to, Darwin. In 1961, John Whitcomb and Henry Morris had published *The Genesis Flood*, an effort to prove that all fossils, and all geology, are the result of the forty-day inundation of which Noah's family and all the creatures he could fit into his ark were the only survivors.[35] The Earth is less than 10,000 years old; all living species are descended from the animals (*Genesis* is silent as to the survival of plants) that Noah saved in his boat; Darwin's theory and all the purported evidence supporting it is abjectly mistaken. Especially, all the techniques that scientists have developed for dating things, from tree rings to carbon-14 to potassium-40, are unreliable.[36] The conclusions are inescapable. "It is quite plain that the processes used by God in creation were utterly different from the processes which now operate in the universe! . . . We must accept either the current theories of paleontology, with an inconceivably vast time-scale for fossils before the appearance of man on the earth, or we must accept the order of events as set forth so clearly in the Word of God."[37]

Those last statements, of course, give the game away. *The Genesis Flood* is obviously a work of religion, not science. Nevertheless, as a widely read and influential exegesis at the time of the *Epperson* decision, it suggested a strategy for creationists to pursue in the future. If they could tone down the explicit references to the *Bible* and put some serious thought into squaring the story of Noah with geological evidence, perhaps they could fashion an alternative science that would pass First Amendment muster.

To put it mildly, the task has proven difficult. Allegedly scientific discussions by creationists keep degenerating into declarations such as this one, from a book published in 2006: "[R]ight down to quarks, God fashioned all these things from an incomprehensible source completely beyond the reach of science."[38] Despite their evident failures, however, creationists keep working at their task of inventing a new science that is consistent with old beliefs. More importantly, they keep declaring that they have succeeded.

As many observers have pointed out, the changes creationist ideas have undergone since the *Epperson* decision follow an almost classic Darwinian pattern. When their intellectual environment has changed, creationists have adapted their self-labeling and sales strategy to the new circumstances. The history of creationism is, then, a history of intellectual evolution. The irony amuses pro-science advocates, but it does not give pause to creationists, who carry on.

Thus, after a spate of books had appeared during the 1970s, following the trail blazed by Whitcomb and Morris, but being slightly more circumspect in emphasizing the *science* in "creation science" and de-emphasizing the *creation*, the Arkansas legislature tried again. In March 1981, it passed the "Balanced Treatment for Creation-Science and Evolution-Science Act," mandating that the public schools should teach both theories in biology classes.[39] (It may be of interest to students of American political history that Governor Bill Clinton signed the bill into law.) The American Civil Liberties Union quickly filed suit, and, in a remarkably efficient procedure for the federal court system, the case was heard by Judge William Overton in December.

The strategy for the creationists was simply to follow the line they had been developing for two decades. There is a Darwinian science and a creationist science; it is only fair to present both sides in public schools.

The strategy for the ACLU, joined by a number of religious/clerical organizations, was two-part. First, in a sort of preemptive ideological strike, its attorney emphasized the position represented by the top two quadrants in Table 3.1. Religion and science are entirely compatible; to teach evolution is not to undermine Christianity. The lead lawyer for the plaintiffs, Robert Cearley, opened by assuring the judge: "We make in this case no challenge to any religious belief."[40] Cearley put on the stand a number of Arkansas religious leaders to testify that not only did they see no challenge to faith from science but that they wholeheartedly supported the separation of church and state. Then he put on a number of distinguished evolutionary biologists to say the same thing. Second, the ACLU called on a number of philosophers and theologians to testify that, in their expert opinion, creation science is religion, not science. It violates all the philosophical requirements of science and embraces the philosophical requirements of faith.

It was a short case that did not provide much intellectual challenge. Judge Overton ruled that the two-sciences law was "a religious crusade, coupled with a desire to conceal" and that "the evidence is overwhelming that both the purpose and effect of Act 590 is the advancement of religion in the public schools."[41] Again, science education was saved from the menace of democracy.

But not for long. The same year as Arkansas passed Law 590, the Louisiana legislature had passed another "Balanced Treatment for Creation-Science and Evolution-Science Act," which contained some provisions not found in its sister state's statute. The Louisiana law, in addition to mandating that if one "science" was taught in public schools, the other must be taught also, had specifically stated that its purpose was to protect the academic freedom of biology teachers who might want to present an unorthodox scientific viewpoint, and had forbidden school boards to discriminate against such teachers.[42] It had also stated that its intention was to protect the freedom of students who might want to avoid indoctrination. Louisiana, therefore, made a strenuous effort to convince everyone that it had a secular purpose in passing the law, and that therefore the statute passed the *Lemon* test.

As in Arkansas, the law was immediately challenged and lost quickly in the lower federal court. But Louisiana, unlike Arkansas, chose to appeal to the United States Supreme Court.

The 7 to 2 majority ruling in *Edwards v. Aguillard*, rendered during the 1987 term, is noteworthy for its lack of innovation. Again, as Justice Brennan delivered the decision, the Justices regarded the Louisiana legislature's announcement of a secular purpose as a sham, an attempt to hide its real purpose of giving an advantage to a theory that "rejects the factual basis of evolution in its entirety [and that] embodies the religious belief that a supernatural creator was responsible for the creation of humankind." Therefore, "the Act furthers religion in violation of the Establishment Clause."[43]

Except for the fact that it continued the Court's secularizing trend, the majority opinion was thus uninteresting and contained nothing of further consequence.

Justice Antonin Scalia's dissent, however, was momentous. It contained a number of the themes, claims, and misunderstandings that have recurred in general creationist writing since 1987, plus a few arguments that pertain specifically to the jurisprudence of evolution. It deserves close attention.

In regard to evolution versus creationism in general:

1. Scalia apparently believes that the modern evolutionary synthesis is a theory of the origin of life.[44] Since all biologists since 1859 have specifically acknowledged that they do not know how life began, it is difficult to see how he could have acquired this idea, except by not paying attention. The misunderstanding enables Scalia to defend the Louisiana legislature for merely wanting to present students with a "fair and balanced presentation of the scientific evidence." Since there is no scientific evidence for the creationists' origin story, accepting Scalia's theory would mean that teachers would have been able to present the first chapter of *Genesis* as a theory, with nothing as an alternative theory. Not surprisingly, pointing out that biologists cannot do what they do not claim to be able to do has become a large part of the text in every creationist tract since 1987.

2. They are both just theories, so what's the big deal? Although his exposition of his position is garbled, Scalia apparently subscribes to the common confusion that a "theory" in science is the same as it is in everyday speech—a mere conjecture, which might or might not be true. Therefore, he endorses the law's stipulation that both creationism and Darwinism be "taught as a theory, rather than proven scientific fact." Then he asserts that it "surpasses understanding" how the Court majority could have come to the conclusion that this balanced treatment would favor religion.

   In point of fact, biologists have practically exhausted themselves writing essays, chapters, and entire books pointing out that in science, "theory" has a somewhat different meaning than it does in everyday speech.[45] A scientific theory is a logically connected system of hypothetical statements about empirical reality from which testable empirical hypotheses can be deduced and which has in fact been supported over many such tests. The theory of evolution, in its modern form, has generated a large number of testable hypotheses over a wide range of biological phenomena and been supported many times. Creationism is a collection of vague, often contradictory, and sometimes nonsensical suppositions that generally is unable to generate empirically testable predictions, and on the few occasions when it has done so, these have not been supported.

   For Scalia to endorse the Louisiana legislature's misunderstanding of the concept of "theory," therefore, is to endorse the conflation of pseudo-science with actual science. Creationists, perhaps encouraged by his dissent, have continued the conflation.

3. Scientists have an "almost religious faith in evolution," and are prepared to conduct pogroms against those who are convinced by the alternative theory. Therefore, it was proper for the Louisiana legislature to seek to protect any hypothetical creationist teachers from discrimination. But Darwinist teachers do not need protection, because they belong to the persecuting elite.
If creationism had a proper standing as an alternative scientific theory, of course, this attitude might by justified. Since it is not, however, Scalia's solicitude for the potential victims of scientific inquisition is another manifestation of the anti-modernist paranoia.

In regard to some specifically legal points:

1. Scalia rejects the *Lemon* test because it relies on the legislation having a secular purpose. It is impossible to base anything on the alleged motives of legislators, he argues, because "the number of possible motivations . . . is not binary, or indeed finite."[46] He then presents a list of possible reasons why a specific legislator might have voted for the Louisiana law for secular rather than religious reasons, including to advance his career by acting as an agent for a religious constituency. The implication of Scalia's argument is that the Court should see a violation of the establishment clause only in a statute that specifically established a state religion. Anything that advanced a religious agenda in public schools, no matter how blatantly, would not be an establishment of religion.
2. There is a tiny number of philosophers, and even some scientists, who reject the Modern Synthesis, or both reject scientific biology and endorse creationism. Some of their affidavits were presented as evidence to the Supreme Court. The majority of the justices, apparently not wanting to go through the process of establishing, once again, that creationism was not a viable alternative to evolutionary biology, passed over the question of which theory was right and made their ruling on the basis of the intention of the law. But Scalia played "gotcha!" in his dissent, pointing out that the Court record contains "ample uncontradicted testimony that 'creation science' is a body of scientific knowledge rather than revealed belief."[47] He inferred that creationism has at least as much scientific standing as Darwinism and that therefore the effort to give it as much presence in the public schools was the result of a secular rather than a religious motive. His argument was sophistry, but that also is part of the jurisprudence of evolution.

## The Arrival of *Homo Designus*

After the defeats in *McLean v. Arkansas* and *Edwards v. Aguillard*, creationism evolved again. The "Intelligent Design" movement emerged in the early 1990s. In Chapter

Three I analyzed three important arguments often employed by the ID partisans, and in Chapter Six I will discuss ID as a political movement. Here, I will examine its relation to Constitutional jurisprudence.

During the fall of 2004, the school board of Dover, Pennsylvania changed its curriculum policy for the ninth-grade biology classes being taught in the public schools under its authority. In addition to offering the ID textbook *Of Pandas and People* as alternative reading, the board required that this statement be read in class before the section of biology dealing with evolution:

> The Pennsylvania Academic Standards require students to learn about Darwin's Theory of Evolution and eventually to take a standardized test of which evolution is a part.
>
> Because Darwin's Theory is a theory, it is still being tested as new evidence is discovered. The Theory is not a fact. Gaps in the Theory exist for which there is no evidence. A theory is defined as a well-tested explanation that unifies a broad range of observation.
>
> Intelligent Design is an explanation of the origin of life that differs from Darwin's view. The reference book, *Of Pandas and People*, is available for students who might be interested in gaining an understanding of what Intelligent Design actually involves.
>
> With respect to any theory, students are encouraged to keep an open mind. The school leaves the discussion of the Origins of Life to individual students and their families. As a Standards-driven district, class instruction focuses upon preparing students to achieve proficiency in Standards-based assessments.[48]

Several parents immediately filed suit in federal court, which, in another miracle of governmental efficiency, heard the case almost immediately. As with previous First Amendment cases, the plaintiffs were represented by the legal counsel supplied by the American Civil Liberties Union. The Dover District was represented by counsel from the Thomas More Law Center, an Ann Arbor, Michigan not-for-profit law firm. According to its website, its lawyers offer their services as a "sword and shield for people of faith" in defense of "the religious freedom of Christians [and] time-honored family values." In case anyone might miss the point, it describes itself as "the Christian response to the ACLU."[49]

Testimony in the case of *Tammy Kitzmiller et al v. Dover Area School District et al* tracked the testimony offered in *McLean v. Arkansas* almost exactly. As they had in the earlier case, witnesses and lawyers for the defendants argued that Intelligent Design, while it might have been motivated by religious concerns, was in fact a scientific doctrine and that, therefore, the school board was just trying to offer alternative biological options to students. Among the witnesses was Michael Behe, whose views I have discussed in Chapter Three. Dover school board members testified that their motivation in requiring teachers to read the disclaimer aloud was to offer another scientific choice to the students, not to urge a religious perspective upon them. (In view of the way the Thomas More Law Center describes itself, someone with a taste for exposing contradictions might have asked why a group of lawyers who advertise themselves as pursuing specifically religious cases was putting on the stand witnesses who specifically swore that their motives were not religious. But apparently no one thought to do so.)

Witnesses and lawyers for the plaintiffs argued that ID was thinly disguised Christian creationism, which the school board was fraudulently presenting as a bona fide scientific alternative to the theory of natural selection. Among the witnesses was Kenneth Miller, a Brown University biologist who has been very active in opposing ID in several venues, some of whose views I have cited in previous chapters.[50]

People who write about the Dover trial invariably point out that the presiding federal judge, John E. Jones III, was a conservative Republican who had been appointed to the bench in 2002 by President George W. Bush. Jones's partisan membership is presumably of interest because, given the coalition of political forces I will discuss in Chapter Six, it might be expected that he would have been sympathetic to the conservative Christian point of view.

Perhaps confounding expectations, however, Judge Jones's evaluation of Intelligent Design was nearly indistinguishable in tone from Judge Overton's evaluation of creation science more than two decades earlier.[51] (It is especially interesting because Overton was a Democrat appointed by President Jimmy Carter.) Although his opinion covered 139 pages, Jones's finding conveyed considerable impatience with ID's claim to being scientific competition to the theory of natural selection.

"The overwhelming evidence at trial established that ID is a religious view, a mere re-labeling of creationism, and not a scientific theory," he wrote. The fact that the Dover school board's disclaimer was intended to be read to students before class not only disavowed the normal scientific curriculum but recommended a non-scientific alternative implying board approval of religious principles. Even though the disclaimer was presented as simply a reminder that there were other possible choices of reading, to offer ID as a legitimate alternative to Darwinism was to violate the *Lemon* test by deliberately advancing religion. Moreover, the claim of the ID proponents that they were simply trying to teach about a controversy "is at best disingenuous, and at worst a canard. The goal of the IDM [Intelligent Design Movement] is not to encourage critical thought, but to foment a revolution which would supplant evolutionary theory with ID." Therefore, the school board's actions constituted an attempt to establish religion, thus violating the First Amendment.

The Dover school board did not appeal Judge Jones's ruling, so this did not become a question for the Supreme Court.

There are two non-judicial points of interest in regard to the Dover trial. The first, an action of the voters in the town, took place even before Judge Jones rendered his verdict. Four days after the end of the trial, all eight incumbents of the school board were ousted in an election by challengers who had pledged themselves to reverse the evolution policy.[52] The trial had drawn national media attention, and so the board challengers had been interviewed on network television news programs, thus permitting non-residents of Dover to hear their rationales for running (and, presumably, the citizens' rationales in voting for them). The challengers made it clear that they were good, believing Christians, and not the agents of a humanist conspiracy to drain morality out of the schools. They simply thought that science, not religion, belonged in public school science classes. They were situated, in other words, within the cadre of citizens in the upper left-hand quadrant of Table 3.1, those people of faith who refused to buy the argument that science and religion are incompatible. They belonged, in another rendering of political space, to the moderate middle, the large portion of the American population who believe that we can live in a modernist society and still pursue spiritual lives, the group that exists in a world of "faith without fanaticism," as Putnam and Campbell put it in *American Grace*.[53]

The second point of interest involves two public statements by Pat Robertson on his conservative Christian television program, *The 700 Club*. Shortly after the Dover election, he spoke to his audience of millions: "I'd like to say to the good citizens of Dover, if there is a disaster in your area, don't turn to God. You just rejected him from your city, and don't wonder why he hasn't helped you when problems begin." A few days later, he expanded on the thought by suggesting to the citizens, "If they have future problems in Dover, I recommend they call on Charles Darwin.... Maybe he can help them."[54]

Most of the media commentary that followed these two analyses of the Dover elections focused on their shocking mean-spiritedness and their medieval belief that God might punish election outcomes with natural disasters. More importantly for our purposes, however, was the fact that the statements underscored the deceptive nature of the Intelligent Design arguments. If ID were science, why would God punish its lack of appearance in a public school curriculum? As Jeff Brown, a former Dover school board member, observed to journalist Edward Humes, "According to sworn testimony, intelligent design has nothing to do with God. Then Pat Robertson says if you don't support it, God will hate you. These clowns want it both ways."[55]

## Conclusion

For all its frequent ambiguities and occasional reversals of course, the American jurisprudence of evolution has been mainly clear and consistent since the 1960s. Within the irreconcilable contradiction that forms so much of Constitutional jurisprudence, the justices of the Supreme Court have chosen Hamilton's perspective, albeit in the prevention of a different type of policy. The people (although this generalization seems not to apply to the citizens of Dover) are not to be trusted with making law at the intersection of government and religion. The First Amendment is to be used as a bulwark against the menacing tendency of democratic majorities to try to impose their religious preferences on everybody else.

The Constitutional doctrine that has arisen from this ideological choice is, in the main, easy to describe. Under the establishment-of-religion and free-exercise clauses, government may not advance religion or allow majorities to impose their religious convictions on those who might have other beliefs. Public schools are agents of government. Modern evolutionary biology is part of the secular scientific project; it is not a separate religion on its own, and its methodology does not derive from the equivalent of a faith-based ontology. Alternatives to scientific biology, whether of the unabashed creationist, or scientific-creationist, or Intelligent Design varieties, are religion rather than science. Any attempt to substitute creationism for scientific biology in the public schools or mandate that the one be taught alongside the other violates the First Amendment and is therefore forbidden.

The previous qualifier "since the 1960s," however, is significant. As I have discussed in Chapter Three, and as I will argue in Chapter Six, a large part of the modern American conservative agenda is to undo the 1960s, in electoral politics, legislation, and jurisprudence. This conservative agenda has been partially realized at the levels of politics and legislation, although it has not yet triumphed in the area of Constitutional interpretation. Such efforts to advance it as Justice Scalia's dissent in *Edwards v. Aguillard*, however, show that the political forces that are powerful in the rest of society are not absent on the Supreme Court. Another set of elections, another set of presidential appointments to the Court and the jurisprudence of evolution that has been settled Constitutional interpretation over the last half-century could alter radically.

To use a metaphor, then, the jurisprudence of evolution, as it has been adapted since the 1960s, will certainly need to continue to evolve. In its present form, it might become extinct. Its environment is hostile.

## Notes

1 Hamilton, Alexander, *Federalist #22* in Hamilton, John Jay, and James Madison, *The Federalist Papers* (New York: Modern Library, 1937), 141.
2 Hamilton, #6 in ibid., 27.
3 Hamilton, #78 in ibid., 508.
4 Martens, Allison M., "Reconsidering Judicial Supremacy: From the Counter-Majoritarian Difficulty to Constitutional Transformations," *Perspectives on Politics*, 5, no. 3 (September 2007), 447–459; Wenzel, Nikolai G., "Judicial Review and Constitutional

Maintenance: John Marshall, Hans Kelsen, and the Popular Will," *PS: Political Science and Politics*, 46, no. 3 (July 2013), 591–598.
5. Lever, Annabelle, "Democracy and Judicial Review: Are They Really Compatible?" *Perspectives on Politics*, 7, no. 4 (December 2009), 805–822.
6. Richards, Mark J. and Herbert M. Kritzer, "Jurisprudential Regimes in Supreme Court Decision Making," *American Political Science Review*, 96, no. 2 (June 2002), 305–320.
7. Kent, James, *Commentaries on American Law*, 1826.
8. Bork, Robert H., *The Tempting of America: The Political Seduction of the Law* (New York: Simon and Schuster, 1990), 7.
9. Ibid., 131, 144–146.
10. Breyer, Stephen, *Active Liberty: Interpreting Our Democratic Constitution* (New York: Random House, 2005), 99.
11. White, G. Edward, "The Path of American Jurisprudence," in *Patterns of American Legal Thought* (Indianapolis: Bobbs-Merrill, 1978), 18–73.
12. I explore this topic in much more detail in *The Paradox of Democratic Capitalism: Politics and Economics in American Thought* (Baltimore: Johns Hopkins University Press, 2006), 118–125.
13. *Engel v. Vitale*, 370 U.S. 421–432 (1962).
14. Habibi, Shahrzad, *God, School, and the Fourth R: Religion Policy in Texas Public Schools*, unpublished undergraduate honors thesis, University of Texas at Austin, 2005; Greenawalt, Kent, *Does God Belong in the Public Schools?* (Princeton, N.J.: Princeton University Press, 2005), 37.
15. This discussion of *Reynolds v. United States*, including the quotation from Jefferson, is based on material in Hamilton, Marci A., *God vs. the Gavel: Religion and the Rule of Law* (Cambridge: Cambridge University Press, 2005), 207, 211–212.
16. *West Virginia State Bd. of Educ. v. Barnette*, 319 U.S. 624, 63 S.Ct. 1178, 87 L. Ed. 1628, (1943), in Steven H. Shiffrin and Jesse H. Choper, eds., *The First Amendment: Cases—Comments—Questions* (St. Paul, Minn.: West Publishing Company, 1991), 708.
17. Witte, John, Jr., and Joel A. Nichols, *Religion and the American Constitutional Experiment*, 3rd ed. (Boulder, Colo.: Westview Press, 2011), 159.
18. *Lyng v. Northwest Indian Cemetery Protective Association*, 485 U.S. 439, 447, 451–452 (1988)
19. Hamilton, *God vs. the Gavel*, op. cit., 214–215.
20. Friedman, Lawrence M., *American Law in the 20th Century* (New Haven, Conn.: Yale University Press, 2002), 509; Greenawalt, *Does God Belong?* op. cit., 21.
21. Hamilton, *God vs. the Gavel*, op. cit., 218.
22. Ibid., 209, 216–223; Witte and Nichols, *Religion and the American Constitutional Experiment*, op. cit., 154–155.
23. *Mozert v. Hawkins County Board of Education*, 827 F.2d 1058 (6th Cir. 1987), cert. denied, 484 U.S. 1066 (1988).
24. Jefferson quoted in Fraser, James W., *Between Church and State: Religion and Public Education in a Multicultural America* (New York: St. Martin's Press, 1999), 19.
25. *Everson v. Board of Educ.*, 330 U.S. 1, 67 S. Ct. 504, 91 L.Ed. 711(1947); *Zelman v. Simmons-Harris*, 536 U.S. 639 (2002).
26. *McCreary County v. ACLU*, 545 U.S. 844 (2005).
27. *Texas Monthly v. Bullock*, 489 U.S. (1989).
28. *Lemon v. Kurtzman*, 403 U.S. 602, 612–14 (1973).

29 Witte and Nichols, *Religion and the American Constitutional Experiment*, op. cit., 173.
30 Arkansas statute quoted in Shiffrin and Choper, *The First Amendment*, op. cit., 644.
31 Background information on *Epperson v. Arkansas* from Wikipedia website.
32 *Epperson v. Arkansas*, 393 U.S. 97, 89 S. Ct. 266, 21 L. Ed. 2nd 228 (1968).
33 Ibid.
34 Fraser, *Between Church and State*, op. cit., 161.
35 Whitcomb, John C. and Henry M. Morris, *The Genesis Flood: The Biblical Record and Its Scientific Implications* (Phillipsburg, N.J.: Presbyterian and Reformed Publishing Company, 1961).
36 Ibid., 43, 343.
37 Ibid., 223, 473.
38 Poppe, Kenneth, *Reclaiming Science from Darwinism: A Clear Understanding of Creation, Evolution, and Intelligent Design* (Eugene, Ore.: Harvest House Publishers, 2006), 279.
39 Gilkey, Langdon, *Creationism on Trial: Evolution and God at Little Rock* (Charlottesville: University Press of Virginia, 1985), 268; unless otherwise noted, all information on the *McLean v. Arkansas* trial is from this book.
40 Ibid., 81.
41 Ibid., 275, 273.
42 Shiffrin and Choper, *The First Amendment*, op. cit., 645.
43 *Edwards v. Aguillard*, 482 U.S. 578, 107 S.Ct. 2573, 96 L. Ed. 2d 510 (1987).
44 See Scalia's statement that the Louisiana law left students "free to decide for themselves how life began," the dissent reproduced in Shiffrin and Choper, *The First Amendment*, op. cit., 646.
45 Futuyma, Douglas J., *Science on Trial: The Case for Evolution* (Sunderland, Mass.: Sinauer Associates, 1982), 167–169; Miller, Kenneth R., *Only a Theory: Evolution and the Battle for America's Soul* (New York: Penguin, 2008), 1–12; Coyne, Jerry A., *Why Evolution Is True* (New York: Viking, 2009), 14–19; Dawkins, Richard, *The Greatest Show on Earth: The Evidence for Evolution* (New York: Free Press, 2009).
46 Shiffrin and Choper, *The First Amendment*, op. cit., 648.
47 Ibid., 647.
48 Quoted in Humes, Edward, *Monkey Girl: Evolution, Education, Religion, and the Battle for America's Soul* (New York: HarperCollins, 2007), 103.
49 www.thomasmore.org/about/about-the-thomas-more-law-center; accessed June 14, 2013.
50 Humes, *Monkey Girl*, op. cit., 297–306; Miller, *Only a Theory*, op. cit., 103–104, 208, 210.
51 *Tammy Kitzmiller et al, v. Dover Area School District et al.*, 400 F.Supp.2d 707 (2005).
52 Humes, *Monkey Girl*, op. cit., 330–331.
53 Putnam, Robert D. and David E. Campbell, *American Grace: How Religion Divides and Unites Us* (New York: Simon and Schuster, 2010), 547.
54 Robertson quoted in Humes, *Monkey Girl*, op. cit., 330.
55 Brown quoted in ibid., 330.

# 6
# EVOLUTION AND THE PARTY BATTLE

In April of 1802, discouraged by the drubbing his Federalists had received from Jefferson's Democrats in the previous election, Alexander Hamilton wrote a letter to his fellow partisan James Bayard proposing a new strategy for the coming fall campaign. Our enemies, he suggested, have won the support of the masses by "courting the strongest and most active passion of the human heart, *vanity*." By this statement he meant that the Jeffersonians had flattered the self-concept of ordinary Americans by talking continually of equality, thus beguiling them into thinking that if they voted Democratic, they might achieve social and economic parity with the wealthier classes, to whom the Federalists appealed, at least in the Northern states. Hamilton feared that unless the Federalists could "contrive to take hold of . . . some strong feeling of the human mind, we shall in vain calculate upon any substantial or durable result."[1]

Since the Democrats seemed to have monopolized appeals to class envy, Hamilton suggested that the Federalists should start cultivating the other dominant emotion in society: religion. What was needed was a coalition between faith and entrepreneurial capitalism. The Federalists, he proposed, should set up a "Christian welfare program," which would function for the dual purpose of providing charity and vocational training to the members of the lower classes, especially immigrants, and of convincing religious believers to vote for "fit men," that is, Federalist candidates, in future elections.[2]

In general outline, Hamilton's suggestion was implemented, but not for 178 years. Why it took so long for the advocates of business ascendancy, and the advocates of Christian conservatism, to combine for the purpose of winning elections is a story that speaks to the heart of American political history. Whether today's coalition will eventually succeed in reversing policies toward the teaching of evolution in public schools will depend both on the strategies the modernist coalition adopts to combat it and on whether its leaders can succeed in managing its flagrant internal contradiction.

## The Mind of the Conservative

For as long as there has been realistic thinking about politics, observers have suspected that there are relatively stable patterns of thought and emotion in regard to the general direction of government and policy. In particular, two iconic ideologies—conservatism and, depending upon the society and era, either liberalism or progressivism—are of interest to anybody who studies American democracy. These two tendencies do not exhaust the possible orientations to government activity, but they have organized much of the debate over who should control government and what it should do, at least as far back as the 1930s.

Use of the terms tended to be loose and journalistic until after World War II, when investigators first began to give the two ideologies empirical content. In 1950 appeared *The Authoritarian Personality*, an attempt by several scholars much influenced by German social theory to combine the paradigms of Marx and Freud into a comprehensive explanation of the sorts of personality traits that had resulted in the triumph of Naziism in Germany, and that might do so in the United States.

Theodore Adorno and his colleagues were leftist in their own political leanings and were convinced that capitalism somehow translated into brutal politics. In their scholarly transmutation of this attitude, they claimed that living in a capitalist social climate caused people to have "status anxiety," which induced them to raise their children with a maximum of physical discipline and a minimum of love. The children, when they matured, attempted to deal with their own emotional damage by treating those below them on the social scale with contempt and brutality and glorifying the authority of those above them. As a result, they were likely to be persuaded by the propaganda of various conservative ideologies, including, at the extreme, fascism. Put simply (and not in the phrasing that they themselves used), Adorno and his colleagues argued that fascism was a mass mental illness caused by capitalism. Analyzing many surveys and psychological tests supplying quantitative evidence, they reasoned from their data that American conservatives were highly susceptible to authoritarian, quasi-fascist appeals.[3]

*The Authoritarian Personality* was enormously influential in subsequent decades and became required reading for anyone in graduate school in the social sciences. Furthermore, it was soon followed by other survey research and psychological studies, each applying somewhat different definitions of "conservatism" and employing varying (non-Marxist, non-Freudian) conceptual frameworks. Nevertheless, most of these investigations came to the conclusion that there was something wrong with the character and ideas of conservative Americans. In 1958, for example, Herbert McClosky summarized his survey research into American political ideologies, divided into four categories (extreme liberal, liberal, conservative, extreme conservative) thusly:

> Of the four liberal-conservative classifications, the extreme conservatives are easily the most hostile and suspicious, the most rigid and compulsive, the quickest to condemn others for their imperfections or weaknesses, the most intolerant, the most easily moved to scorn and disappointment in others, the most inflexible and unyielding in their perceptions and judgments.[4]

This research tradition of elaborating the personality deficiencies of political conservatives, employing ever more sophisticated methodologies and applying ever more discerning models of cognition, has continued into the twenty-first century.[5] It has, however, attracted serious, sustained criticism.[6]

The first problem with the "conservatives are dangerous to democracy" scholarship is that it relies on a central concept that is both ambiguous and politically loaded. As to ambiguity—What is conservatism? Is it an attitude toward change? An attitude toward authority? An attitude toward equality? A set of prejudices toward various ethnic and religious outsiders? A set of interrelated ideas about appropriate public policy? And, whatever it is, does it have a stable meaning across time and across cultures? Critics argue that "conservatism" is an historically unstable concept and that therefore it is folly to believe that it covers a set of values or policy positions.[7]

As to political loading: The instability of the concepts "liberal" and "conservative" in American politics is well illustrated by the career of "affirmative action." In 1963, someone who believed that African Americans should be treated as individuals rather than as members of a group, and therefore supported various national efforts to remove state restrictions on their civil rights, was called a liberal. Half a century later, someone who holds exactly the same beliefs, and therefore opposes the policy that, for example, treats African Americans as members of a group rather than as individuals in university admissions, is called a conservative. A policy position that can reverse its ideological membership in one human lifetime is a poor candidate for measurement as a permanent cognitive component. The fact that some investigators persist in considering it such a component raises the suspicion that they are ideological liberals whose work is more polemical than scholarly.

The second criticism of the anti-conservative research is that it is structured so as to ignore authoritarian tendencies on the left of the political spectrum. After the hubbub over the book by Adorno et al., psychologist Milton Rokeach and his colleagues performed a series of experiments and surveys with respondents in both the United States and Britain. Using some of the same measures as had the authors of *The Authoritarian Personality* (whose F-Scale, supposedly measuring proto-fascistic personality tendencies, had instantly become a standard tool in all research in this area), they teased out tendencies to dogmatism, opinionization, and party-line thinking.[8] When questionnaires were reworded to make them ideology-neutral, Rokeach discovered that leftists too could be intolerant and dogmatic. For example, he found that in Britain, members of the Communist Party scored the highest of all groups in dogmatism, despite scoring the lowest of all groups on the F-Scale.[9] In other words, he demonstrated that while Adorno may have identified a right-wing authoritarian personality type, he had missed the equally significant existence of left-wing authoritarian personalities.

Rokeach's conclusion was that there were open-minded personalities, relatively hospitable to new ideas, relatively tolerant of people who disagreed with their own positions and values, and relatively less trusting of authority. Conversely, there were closed-minded personalities, who were relatively hostile to new ideas, intolerant of disagreement and punitive toward people whose ideas differed from their own, and blindly obedient to authority.[10]

The sensible conclusion to be endorsed after reading such research is that partisans of all belief systems contain some representatives of both open-minded and closed-minded people. There are closed-minded liberals and closed-minded conservatives, closed-minded lovers of Italian cooking and closed-minded baseball fans. There might even be some closed-minded scientists. Nevertheless, it is entirely possible that some ideologies and some organizations contain different mixes of the two personality types. Rokeach surveyed Catholics, Protestants, and non-believers in the United States, reporting that the Catholic respondents contained the highest percentage of closed-minded people, while the non-believers contained the lowest, with the Protestants in the middle.[11] He did not break down the Protestants in his sample into fundamentalists and mainstream categories, or evangelical and non-evangelical, so his analysis cannot be applied directly to questions of the politics of evolution. But subsequent researchers have investigated the point at issue.

Unfortunately for clear exposition, the various studies that have been done in the general area of dogmatic belief tend to use slightly different characterizations of important concepts. That is, sometimes we are studying tolerance, sometimes closed-mindedness, sometimes an absolutist approach to moral issues, and so on. Sometimes, the groups we are studying are fundamentalists, sometimes evangelicals, sometimes conservative Protestants, and so forth. It is therefore impossible to summarize with complete precision what social investigators have discovered about ideology, religion, science, and politics in America. Nevertheless, assuming that the concepts are roughly equivalent and that the groups overlap in membership, some findings emerge strongly from several decades of social science effort.

It is probably not true that political conservatives, however defined, are by nature dogmatic and intolerant. But it does seem to be true that fundamentalist Christians are, in general, "morally absolutist" in their approach to questions of right and wrong and commonly view such concessions to human weakness as pornography as moral contaminants, "viewing the presence of sin in society as leading to social decline," in the words of Sherkat and Ellison.[12] In addition, fundamentalists are intolerant of social outsiders such as homosexuals, communists, and atheists. That is, they are willing to deprive those individuals of such civil rights as the liberty to give a speech in public.[13] Of course, given their view that the *Bible* offers a literally true narration of the history of life, they reject any scientific theory of evolutionary development.[14] Although no scholar appears to have put the facts together in quite this way, it is thus obvious that fundamentalists are convinced that belief in evolution is actually sinful.

Further, Southerners as a geographic group share many characteristics with fundamentalists. They are, in general, an intolerant and closed-minded bunch. Part of the explanation for this ideological overlap lies in the fact that fundamentalists tend to be concentrated in the states of the old Confederacy and the border states of Oklahoma, Missouri, and Arkansas, becoming rarer as one travels northeast and northwest.[15] Nevertheless, research has shown that Southern intolerance is a force independent of religious conviction.[16] That is, Southerners, whether fundamentalist Protestant or not, tend to be less tolerant than non-Southerners, and fundamentalists, no matter where they live, tend to be less tolerant that non-fundamentalists.

Additionally, there is an interaction between politics and religion. Over time, Americans have been slowly sorting themselves into a religious–politically conservative–Southern–Republican group and a secular–politically liberal–non-Southern–Democratic group. Given the fact that all these categories have porous boundaries, permit many instances of cross-pressures, and vary with the issues that are prominent at any one time, the generalizations miss many individual exceptions and ideological nuances. Furthermore, participation in the democratic political process serves to moderate some of the intolerance of conservative Christians—activists tend to become more willing to compromise and to concede the good intentions of those with whom they disagree.[17] Nevertheless, as overall characterizations, the portrait of a large bloc of religious conservatives agitating to impose their theological views on everyone else captures an American historical trend.

Finally, an important qualification to the above three summary points is the difference between African Americans, on the one hand, and Hispanics and non-Hispanic whites ("Anglos"), on the other. The movement of people with conservative Christian views toward the Republican party applies only to Anglos and

Hispanics. Fundamentalist blacks, like non-fundamentalist blacks, have remained steadfast in their Democratic partisanship over time.[18] Furthermore, even when religious denomination is controlled, they tend to be more tolerant of moral transgressions than Anglos, which suggests that they are more open-minded in general.[19]

Taken together, these various facts and trends suggest that there is indeed a "culture war"[20] that has been ongoing in America for some decades but that the two sides in the conflict are neither rigidly defined nor exclusive. At one end of the conflict continuum are white Southern fundamentalist conservative Republicans, who can be expected to support the idea of using state power to impose their own version of religiously derived morality on everyone else. Such a position would include teaching creationism rather than scientific theory in public schools. At the other end of the spectrum would be a group of less easily defined Americans, who would tend to be less Anglo, non-Southern, more secular, more inclined to vote Democratic, and, in general, both more accepting of non-mainstream behavior and more inclined to accept evolution, and only evolution, as a fit subject for children's education. Between these two polar groups there would be individuals who are cross-pressured, say, Southern atheists or Northern evangelicals. Part of the effort of culture warriors would be expended trying to persuade members of the conflicted middle to join or at least acquiesce to the dominance of their side.

The third major criticism of the view that conservatives are bad for democracy lies in the fact that modern democratic politics has several issue dimensions. There is an economic issue dimension, in which conservatism lies at the pole at which someone is pro-business and anti-labor, opposes regulation of economic activity, and opposes government redistribution of wealth. There is a social dimension, most relevant to this book, in which a conservative lies at the pole at which he or she endorses traditional religious morality and ethnic prestige hierarchies. There is a foreign-policy dimension, in which a conservative lies at the pole of hawkishness/pro-intervention abroad. Whether two or all three of these dimensions line up together in one person's mind or whether a person is positioned along each dimension at a different point is a question for investigation for any given person and any given era. Accounts of "conservative" ideology that leave out the constantly changing, multidimensional structure of political beliefs are bound to be misleading.

In his letter to Bayard, Alexander Hamilton was suggesting that the Federalists needed to bring together people whom today we would call conservative on the social dimension with the conservatives on the economic dimension who were already supporting the party. Why that alignment of opinion and partisanship did not happen then and why it did happen much later are questions of historical causation that are surprisingly easy to answer. The question that is hard to answer is what the present structure of opinion and party means for the future of the teaching of evolution in the public schools.

## Slouching toward Disequilibrium

American party politics has always teetered on the brink of a collapse into chaos. The reason for its loud, jumbled quality has been clarified by various kinds of theoretical thinking over the past half-century.

It is possible that national politics would, during some eras, consist of a single issue dimension. That is, at some times there might be one question that is overwhelmingly important to a large majority of the citizens. Such a question might be, for example, "Shall we allow the extension of slavery into the territories?" or "Shall we begin to regulate capitalistic investment decisions?" or "Shall we limit carbon emissions in order to try to stop global warming?" Under such conditions of a single issue dimension, theoretical work stretching back to the 1930s provides a clear portrait of equilibrium. If there are two parties, then there is one point among the various positions that a party or candidate might take that is in everyone's interest to endorse, and nobody's interest to depart from. In the case of one-dimensional democratic politics, and two-party politics, the equilibrium point is at the center of the continuum. Both parties will converge to the median position. If there are more than two parties, the question is more complicated, but the principles at work are, in general, the same.[21]

In an issue space with two or more dimensions, however, there is no equilibrium point. That is, no matter what combination of issue positions a candidate or a party takes, there is another combination that can beat it—that can attract more votes. Scholars have published many articles containing sophisticated mathematics that elaborate on this point. I will not draw a graph or print an equation here, because the principle can be expressed fairly clearly in words. In an abstract world in which citizens have perfect information but no fixed loyalties, in any given election they will simply vote for the candidate who best represents their opinions, and as a result will be constantly changing their partisan support. In the same abstract world in which candidates have no motivation except victory at the polls, they will be constantly shifting their positions to outflank other candidates, and avoid being outflanked.[22]

In the real world, politics is more stable, for a variety of reasons. Citizens do not have the inclination or patience to be always participating in a gigantic game to maximize their policy preferences by choosing to vote for the person or party that best realizes their desire in a given election. They therefore rely upon heuristics (rules of thumb), the most famous of which is party identification, to allow them to more or less choose the candidate closest to their own preferences, without having to spend all their time informing themselves. Many citizens adopt a party identification as a standing decision in the voting booth. The distribution of party identifications in the electorate tends to change slowly in response to historical events and the turnover of generations. As a result, there tends to be a drag on electoral outcomes, with partisan balances sometimes adjusting over a very long time to policy changes at the national level.

Additionally, candidates are not simply trying to cultivate voters. They have two other constituencies to pacify: party activists and major contributors, or "investors," who typically represent some large economic or social interest in society.[23] The position on the issue spaces of these two groups may be very different from the position of the voters, and from each other. Candidates and parties, therefore, have to perform an acrobatic calculation about how to construct a winning coalition by simultaneously assuring the support of activists and investors while attracting enough voters to win elections.[24] The strategies they employ to do that need not concern us in detail, but involve the use of ambiguity, rhetorical symbolism, and lies.

Finally, because of federalism and the separation of powers, and, in particular, the independence of the courts, electorally victorious parties do not have complete freedom to make policy. Thus, several times in American history, a party has achieved majorities in both houses of Congress and captured the White House, yet been thwarted in its policymaking aspirations, at least for a while. The ability of parties to maneuver within a policy space, and thus reward or tempt various constituencies, is thus retarded by the "institutional friction" of the Constitutional system.[25]

As a consequence of the various types of drag on the system, two-party democratic politics tends to follow a course of a slow-motion fall into disequilibrium, punctuated by occasional periods of stability. As one party establishes a majority governing coalition in Washington, the minority party begins maneuvering to pick off some of its supporters among the voters, activists, and investors by adopting new policy positions in the multidimensional space. This activity is, of course, tricky, because by modifying its rhetoric and policies to attract new supporters, a party risks alienating old supporters. And the governing party is maneuvering also, trying to increase its majority. The result of the constant, disequilibrated competition is that the parties tend to slowly move through the abstract space of actual and potential support, occupying some issue positions for a certain period, then creeping on to some other positions as public opinion, their own electoral fortunes, and the rhetoric of their antagonists change through history. Over time, they might actually swap positions on important policies.

Partly because of the inherent stickiness of political change, and partly because of its accidental association with an issue that was far more important in American politics, evolution did not emerge as a point of contention between the parties until more than a century after Darwin wrote. As the creep of coalition building proceeded over the twentieth century, however, the kaleidoscopic recombination of issues gradually removed both the stickiness and the accidental association. As the project of modernization discussed in Chapter Three went through the kaleidoscope, evolution came into focus as a political issue, and the ghost of Alexander Hamilton began to smile.

## That Old-Time Scholarship

Well into the 1960s, political scientists regarded the politics of the Southern states as being covered by discussions of two subjects: economic class and race. As for economic class, Marx had probably been the most important single influence over the way the generation of scholars coming out of the Depression thought about political conflict. Some very distinguished scholars had been Communist party members at some point in their youth, but even after they rejected Marx's prophecies and moral assumptions, and embraced democracy, they retained his focus on economics as the most important single subject in politics. In 1960 the great political sociologist Seymour Martin Lipset published *Political Man*, his summary of the thinking on the topic up to that time. Elections, he wrote in that book, were the "expression of the democratic class struggle."[26] For decades afterward, graduate students in political science read that book and absorbed its lessons. And although everyone knew that religion was important to people in general and Americans in particular, few scholars in political science thought that it was important enough to include in their theories of voting or their research about political behavior.[27]

They did, however, think that the race problem was important enough to study, especially (although by no means exclusively) in regard to the Southern states. As V.O. Key, Jr. wrote in his monumental and monumentally influential 1949 book, *Southern Politics*, "In its grand outlines the politics of the South revolves around the position of the Negro."[28] But the influence of his "it's all economics" view led Key astray also, and the major, although unstated, theme of his entire 675 pages was that a politics-as-economics struggle in the South (good) was threatening to break through the politics-as-racial-suppression conflict (bad) that dominated the region. The efforts of elites in the South to keep their white citizens' attention focused on maintaining racial segregation might fail, he argued, because "[p]olitics generally comes down, over the long run, into a conflict between those who have and those who have less,"[29] which was merely an anticipatory restatement of Lipset's dictum.

There is no heading for "religion," "faith," "fundamentalism," "evangelism," "Christianity," or "Darwin" in the index of *Southern Politics*. I could find only one brief and indirect reference to the Judeo-Christian tradition in the text, when Key observes, without comment, that one of the planks in the victorious 1938 Texas gubernatorial campaign of W. Lee "Pappy" O'Daniel, was an endorsement of the Ten Commandments.[30] Why a candidate in a Southern state might find that plank useful and how it might have contributed to his victory were not subjects that Key chose to explore. They addressed neither economics nor race.

For most of the twentieth century, everyone of any education knew that evangelical Christianity dominated the South. The term "Bible belt" was a phrase in common usage to identify the region.[31] Stories, novels, and movies set in the South often included fundamentalist preachers. The Scopes trial, and *Baltimore Sun* journalist H.L. Mencken's columns about the evangelical fervor that accompanied it, were part of the historical memory of the country. Yet, when scholars wrote about Southern politics, they typically followed Key's lead and discussed race and economics.

As a consequence, when serious thinkers considered the social-issue quadrants in a multidimensional space, the term "race" monopolized the area. The social politics of the United States and of the South in particular was, as far as political scientists were concerned, a politics of race. It followed, as Key had argued, that if race relations could be defused as a flashpoint of political conflict, then the social policy dimension could be muted, and a more "normal" politics of economic class conflict would reassert itself. Until well into the present era, it did not seem to occur to American scholars that even if the race problem were to magically vanish, the social dimension of American politics would remain as energized and productive of bad feelings as ever, because the rage of fundamentalists at modernizing secularism would be undiminished. Many thinkers have not yet come to terms with this fact. In short, secular politicians, journalists, and scholars have not yet learned to take the politics of evolution seriously.

## Look Away, Dixieland

Because of the situation of the South in national politics, conflicts between that region and the rest of the country tended to revolve around two great issues during the nineteenth century. In the economic dimension, the most important conflict was over the tariff, although the currency question became important after 1873.[32] In the social dimension, the conflict was over black people, first their status as slaves and then, after 1865, their status as citizens. Fervent, evangelical Christianity dominated the white Southern population but was not important in national politics for two reasons. First, the Fourteenth Amendment, which would eventually become a nemesis for state politics, did not rise to relevance until 1925. Therefore, religion, education, and the relationship between the two were entirely state matters.

Second, commerce and industrial capitalism were in an antagonistic relationship to Southern society. Although Senator John C. Calhoun of South Carolina had proposed an alliance between his own Southern, agricultural, slave-based elites and Northern, capitalist, free labor-based elites, political cooperation between the "gentlemen" of each region never materialized.[33] The pre-modern chattel slave system was incompatible with the modernist, free-market capitalist labor system in place in the Northern states.

In the post-Civil War era, conflict over the Northern manufacturers' and bankers' support for the tariff and opposition to expansion of the currency, added to the searing resentment of Southern whites at the results of the Civil War and Reconstruction, created implacable hostility between the two regions.[34] Thus, although the politics of religious conflict sometimes was important in the North, and at the state level, it was only rarely mentioned in national politics.[35]

The election of 1896 further confirmed Southern social conservatives' estrangement from Northern capitalist conservatives. The Democrats, with William Jennings Bryan as their standard bearer, made a determined push to take over Washington on behalf of Southerners, Westerners, farmers, and partisans of inflation of the currency. After their defeat, Democrats retreated into their Southern bastions, content to let economic conservatives run the country as long as they themselves could run the region, which, in practice, meant suppressing the vote of blacks and poor whites.[36] Meanwhile, conservative capitalists occupied themselves with fighting off progressive insurgencies within the Republican party in the North. As the Scopes trial showed, Northern elites might laugh at Mencken's accounts of the antics of the Bible thumpers in the South, but they were not going to try to intervene in Southern local governance. Again, therefore, the political potential of the religious fervor of the South was suppressed beneath other issues.

The System of '96 came apart during the Great Depression of the 1930s, but it did so in a manner that only further deprived religious zeal of its expression. Theoretically, Northern capitalists might have appealed to Southern social conservatives in their campaign to oppose the New Deal's imposition of economic liberalism. But President Franklin Roosevelt, sitting astride a Democratic coalition that included Northern blacks and Southern white racists, kept his rhetorical focus on economic problems. He could not openly support a federal anti-lynching bill, he was once candid enough to explain, because that action would alienate the Southern Democrats in Congress.[37] Despite an occasional local rebellion, therefore, FDR generally commanded the loyalty of Southern white voters, if not always Southern white politicians. The association of religious conservatism with racial conservatism kept Southern fundamentalist anti-modernist rage bottled up within the New Deal coalition until after World War II.

As American history advanced after the war, however, the contradictions within the Democratic coalition began to show. A Cold War between the Western democracies and the Communist world, beginning in 1947, focused much of the attention of successive presidential administrations on a worldwide propaganda battle about the merits of the two systems. The Soviet Union enthusiastically publicized every brutality and injustice visited upon African American citizens in the Southern states, and the United States found itself on the defensive in Third

World countries, especially those in Africa. Meanwhile, white liberals and black activists were growing restive with Southern racism, a discontent that flowered into the Civil Rights movement of the early 1960s. A combination of foreign and domestic policy pressures thus combined to convince the Kennedy and then the Johnson administration to make a push to eliminate the most evident Southern racist state policies.[38] The results were the Civil Rights Act of 1964, the Voting Rights Act of 1965, and various administrative policies to forbid the customary Southern measures to deprive African Americans of the rights of citizenship.

President Lyndon Johnson, if not every social liberal Democrat in the North, knew the risk he was running when he pushed for civil rights legislation. Years later, Johnson's widow recalled that after the news came of the passage of the Voting Rights Act, Johnson walked into the family quarters of the White House, "and there were a few people there ready to do a postmortem on the bill. . . . 'Well,' Lyndon told them, 'I think I just may have handed the solid Democratic South to the Republican party.'"[39]

However, as I have discussed in Chapter Three, the process of disassociating the socially conservative South from the Democratic party had actually begun three years before, with the school-prayer decisions by the Supreme Court. But there is no doubt that the racial issue had a greater impact in the 1960s. Senator Barry Goldwater, the Republican presidential candidate for 1964, had been one of eight Republican senators to vote against that year's Civil Rights Act in Congress, and the pattern of citizen support, with Goldwater taking only the states of the Deep South besides his home state of Arizona, was undoubtedly a reaction to that policy. Moreover, in that election and for years afterward, Southern voters often split their tickets, casting their presidential ballots for Republicans but reelecting Democratic congressional representatives, who had opposed the Civil Rights and Voting Rights Acts almost unanimously.[40]

With the social and economic dimensions severed, and muddied by the reaction to the Vietnam War and then the Watergate scandal, the Republican party crawled around in the two-dimensional issue space in a disjointed and inconsistent manner. Ronald Reagan, however, saw the potential of uniting social and economic (and, incidentally, foreign policy) conservatives. He was the embodiment of a Hamiltonian politician, worshipful of economic entrepreneurs, religious and anti-modernist at the rhetorical level, and even, like Hamilton, a great friend of British civilization. His political career up to 1980 had been an effort to unite the three conservative issue dimensions in his person. In the summer of that year, at the national Republican convention at which he was nominated, and afterward, he made it all come together.

In addition to inscribing the usual Republican themes of lowering taxes and cutting back government regulations on business, the party adopted a platform that, in its social-issue planks, shocked some of its own elected officials. The party dropped its earlier support for the Equal Rights Amendment, endorsed a Constitutional amendment to outlaw abortion, and recommended that opposition to abortion be a prerequisite for a presidential appointment to a federal judgeship.[41] Republican Senator Charles Percy of Illinois charged that the latter suggestion was "the worst plank that has ever been in a platform."[42] But he had already been left behind by his party.

The nomination and the writing of the platform, however, were not all of Reagan's rhetorical appeals to the social conservatives. A month after the convention, Reagan spoke at a conference of evangelicals sponsored by the Religious Roundtable. The organization, and its member churches, could not officially endorse any candidate, because direct political involvement would remove them from tax-exempt status under the U.S. federal income tax laws. It is significant, however, that both incumbent President Jimmy Carter and the major independent candidate, John Anderson ("wallet on the right, heart on the left") had been invited to address the conference and had declined. Each knew that he would pick up no support there. Reagan, however, accepted the invitation with alacrity and did not disappoint the delegates. "I know that you cannot endorse me, but I endorse you," he said, to thunderous applause, thereby cementing an alliance of social and economic conservatives that is still with us.[43]

While Reagan's rhetorical endorsement of what would shortly come to be known as the Religious Right received all the attention it deserved, another of his statements did not get much publicity until later. At a pre-speech press conference, he had warned that "traditional Judeo-Christian values" were in danger, attacked the Supreme Court for expelling God from the classroom, and confessed that if he were shipwrecked and could choose only one book to read for the rest of his life, it would be the *Bible*. With complete consistency, then, he also advocated that the *Genesis* creation story be taught in the public schools as an alternative to the theory of evolution, which he asserted was increasingly being abandoned by scientists.[44]

Research has established that Reagan was elected in 1980 because of national economic problems—the deadly combination of inflation and unemployment that economists term "stagflation"—not because of his stand on social issues.[45] Nevertheless, he was successful in establishing Hamilton's alliance at the level of party coalitions. Slowly and fitfully, the voting of Americans at the congressional and state levels revolved to fit into the alliance of party activists and investors inaugurated in the presidential voting of 1980. By the 2000 national election, the party support of Americans was essentially a mirror image of what it had been in 1896. With the usual exceptions caused by wars, scandals, and personal appeal, the Republican alliance of social and economic conservatives dominated the Southern, Border, Great Plains, and Mountain states. The Democratic alliance of social and economic liberals dominated the Northeastern, Midwestern, and Pacific Rim states.[46] Voting patterns in the South were not quite the same as they had been in 1896 because of legal and demographic changes. Since African Americans and, increasingly, Latinos were now voting, they were able to elect Democratic congressional representatives. Thus, there was no "solid" Republican South, as there had been a solid Democratic South. But there was a South and border states with Mexico dominated by socially conservative Republicans, and pockets of Christian Right activists in many other states.

By the arrival of the George W. Bush administration in 2001, a Southern-based party had fully bought into the Hamilton–Reagan formula and embraced a "mixture of moralism and anti-intellectualism" (in Colleen Shogan's words) that delighted its evangelical constituency.[47] Bush had already endorsed the teaching of "different schools of thought when it comes to the formation of the world.... I mean, religion has been around a lot longer than Darwinism."[48] Meanwhile, all the Senator Percys in the Republican party had retired, been defeated (like Percy in the 1984 election), been converted to the anti-modernist agenda, or were keeping their heads down.

Because of the Supreme Court's insistence on a secular school system, however, Republican administrations could not reimpose Christianity at the national level. The politics of evolution, as the era of culture wars developed, was increasingly fought out in the state legislatures. Because of the modernist opposition of the federal courts, however, the state anti-modernizers needed a strategy, some mixture of rhetorical, electoral, and legislative tactics to try to reimpose the old-time religion in the schools. They found it, although whether it will ultimately be successful is still an open question.

## Intelligently Designed Strategy

Social movements, and the public interest groups that are their practical incarnation, do not simply arise when a group of people becomes disgruntled with public policy. They require leadership. "Policy entrepreneurs," as political scientists like to call them, are those people who organize a small or large group of discontented citizens, give them a common identity and vocabulary, urge them to action and provide them with rules about how to behave, seek to use the media to "frame" issues to the benefit of the group, and create a grand strategy that guides the direction of the group.[49] Beginning in the 1970s, the Christian Right produced a variety of entrepreneurs who inspired interest groups that had an impact on American politics. In regard to the politics of evolution, the most important policy entrepreneur has been Phillip E. Johnson.

Johnson was a law professor at UC Berkeley when, in his late 30s, he underwent a personal crisis, apparently as a result of the failure of his marriage.[50] Retiring from his profession and casting about for something to fill his now meaningless life with purpose, he chanced to read both ethologist Richard Dawkins' *The Blind Watchmaker*, an eloquent description of the logic of evolution, and molecular biologist Michael Denton's *Evolution: A Theory in Crisis*, an insider's attack on the adequacy of natural selection to create new species.[51] He later told an interviewer, "I read these books, and I guess almost immediately I thought, *This is it. This is where it all comes down to, the understanding of creation.*"[52] Like St. Paul, who had undergone a similar dramatic experience on the road to Damascus almost two millennia earlier, Johnson resolved to use his newfound life's goal to organize his fellow Christians to take over the world.

Johnson's first task was intellectual. He wrote *Darwin on Trial*, published in 1991, a prosecuting attorney's accusation against the theory of natural selection rather than a scientific assessment of it. Johnson marshaled every argument he could think of against Darwinism, true or not, fair-minded or not—selection can create small variations but not new species, the fossil record does not support the theory, the molecular evidence contradicts the theory, scientists can't agree on what "evolution" means, blind mutation could not have built complex organisms from amino acids, evolutionists cannot explain how life originally began, the theory is not falsifiable and therefore not scientific, Darwinism is a religion supported by intolerant dogmatists, and, most auspiciously, by only permitting naturalistic (observable, measurable) types of arguments and evidence, modern scientists are ruling out the obvious explanation for the development of life: supernatural miracles. (I have discussed some of these arguments, and particularly the last, in Chapter Three.)[53]

Johnson also took some pains to distance himself from Young Earth Creationists, emphasizing his view that God—and it was God in 1991, the strategy of insisting that life's designer was a vague, unspecifiable entity not yet having occurred to Intelligent Design (ID) partisans—directed the contours of life's flowering over the 3.8 billion years that biologists claimed. One did not need to cling to the literal truth of *Genesis* in order to endorse the idea that all life is a supernatural artifact:

> I am a philosophical theist and a Christian. I believe that a God exists who could create out of nothing if He wanted to do so, but who might have chosen to work through natural evolutionary process instead.[54]

Having constructed the indictment, Johnson then began to organize. Supporters at Southern Methodist University put together a conference on Johnson's book in 1992, inviting various scientists and philosophers they knew to be sympathetic to his point of view. Networking began. The next year Johnson organized another conference, at Pajaro Dunes, California, financed by Howard and Roberta Green Ahmanson. There, although intellectual interchange took place, the emphasis was on fashioning a strategy for spreading the new creationist message, now repackaged as "Intelligent Design."[55] As one of Johnson's most prolific associates, William Dembski, puts it, Johnson is the "fearless leader," the man who is the "ID movement's chief architect and guiding light," the one who "had both the plan and the will" to make the country take notice of ID.[56] In 1996 the new creationist insurgency found a home in Seattle's Discovery Institute, at the Center for the Renewal of Science and Culture.[57] The Discovery Institute had been a free-market conservative think tank, with deep connections within a network of Republican organizations pushing less regulation and lower taxes.[58] Its melding with Intelligent Design symbolizes the culmination of the alliance of economic and social conservatives that Hamilton had recommended and that Reagan had initiated.

Over the course of the two decades since, both the organization's tactics and its vocabulary have changed somewhat in response to historical circumstances, and especially the *Kitzmiller* court decision of 2005 (discussed in Chapter Five). But the basic strategy has remained constant: Do not try to convert scientists and secular intellectuals. Instead, appeal to the "undecided middle ... the majority of Americans, who don't buy the atheistic picture of evolution peddled in all the textbooks."[59] Appeal to Americans' sense of fair play; don't demand that the subject of evolution be dropped from public school science classes, just request that schools be permitted to "teach the controversy."[60] Protest that keeping ID out of the public schools constitutes "viewpoint discrimination," which is the equivalent of religious discrimination, and therefore a violation of the First Amendment.[61] Point out that we are supposed to be a democracy and that large majorities of ordinary people want to teach both Darwinism and creationism, while scientific elitists, upper-class liberals, and unelected judges are uniting to deny them the right to educate their children according to their own values.[62]

And adopt Young Earth Creationist Duane Gish's dictum: Always attack. Never explain.[63]

A survey of what Republicans say and write, in the states where they are dominant, would thus lead us to believe that Intelligent Design is any minute going to be adopted as educational doctrine in those areas of the country where social conservatives seem to run the show. Here, for example, is a plank from the Texas state organization's platform from both 2010 and 2012:

> Realizing that conflict and debate is a proven learning tool in classrooms, we support objective teaching and equal treatment of all sides of scientific theories, including *evolution, Intelligent Design, global warming,* political philosophies, and others. We believe theories of life origins and environmental theories should be taught as challengeable scientific theory subject to change as new data is produced, not scientific law. Teachers and students

should be able to discuss the strengths and weaknesses of these theories openly and without fear of retribution or discrimination of any kind.

Given the fact that Texas, like some other states in the South, Plains, and Mountains, has had overwhelming Republican dominance in its legislature for at least the past decade, it might be assumed that such intentions as those expressed in the party platform would long ago have become public policy. In practical terms, however, the ID movement has had only small and temporary successes in influencing school curricula.

In 1999, the Kansas state board of education, after consulting both Phillip Johnson and Young Earth Creationists, voted to delete references to macroevolution (that is, the evolution of new species, as opposed to in-species variation) and the age of the Earth from the state's science standards. After this action had attracted both national and international ridicule, two of the three creationists on the board who ran for reelection were defeated by pro-evolution moderates in the August 2000 Republican primary. A third was forced to resign from the board after establishing permanent residence out of state. In February 2001, the newly constituted board voted to establish new science standards restoring the previously deleted subjects.[64]

In 2001 U.S. senator Rick Santorum succeeded in convincing most of the other members of the chamber to add a "Sense of the Senate" amendment, written by three ID members, to the No Child Left Behind Act. The language of the amendment had been composed so as to sound as innocuous as possible, yet still encourage creationists to press their case with school boards:

> It is the sense of the Senate that—(1) good science education should prepare students to distinguish the data or testable theories of science from philosophical or religious claims that are made in the name of science, and (2) where biological evolution is taught, the curriculum should help students to understand why this subject generates so much continuing controversy, and should prepare the students to be informed participants in public discussion regarding the subject.[65]

Unaware of the purposes and implications of the language, many liberal Democrats voted for the amendment, and it passed the Senate 91 to 3. But the various organizations and individuals who keep track of this sort of subterfuge caught on to what was happening and raised an outcry. The Santorum amendment was not part of the No Child Left Behind Act when it was reported out of the House–Senate conference committee in December.[66]

In 2004, as reported in Chapter Five, individuals in the ID movement were important advisers to members of the Dover, Pennsylvania school board when it imposed various anti-evolution requirements on high school biology teachers. Again, the ID success was short-lived, as both judicial and electoral disaster soon befell the creationist members of the board.

In Texas, Republicans have held an easy majority on the fifteen-member state board of education since the early 1990s. In the late years of the first decade of the twenty-first century, creationist Republicans occupied seven seats. Yet, a coalition of moderate Republicans and Democrats, maintaining an improbable majority of one, kept voting down ID-inspired curriculum proposals. One member, Chairman Don McLeroy, became nationally famous in the press for his forthright style of expression ("Evolution is hooey!"), and on YouTube for his rants against the teaching of the theory of natural selection.[67] After a while, Republican voters became tired of the series of near-misses with creationism and defeated McLeroy in the 2010 primary.[68] Another creationist, Cynthia Dunbar, chose not to run for reelection and was replaced by a Republican with more moderate views.

The pattern has by now become almost standard. Every year, in many states, especially in the Bible belt, creationist candidates run for boards of education. A few of them are elected, but in general not enough to make a difference. When they are numerous enough to make a difference, members of their own party, in alliance with Democrats, turn against them.

It is the same in the state legislatures. For example, in 2013, Republican legislators in Arizona, Colorado, Indiana, Missouri, Montana, Oklahoma, Texas, and Virginia submitted bills to amend the states' biology curricula.[69] All of these bills bear the imprint of ID philosophy and prose style, requiring "equal treatment of science instruction regarding evolution and intelligent design" or the equivalent and forbidding schools to interfere with instructors who want to teach, or students who want to study, creationism. In seven of the eight states, the bills died in committee. In Oklahoma, the bill was reported out of committee, then defeated on the floor.

Except in Colorado, where the Democrats controlled both houses of the legislature, and perhaps in Virginia, where the two parties were tied in representation in the state senate, the explanation for the failure of these bills cannot be that Republicans did not have enough votes to pass them. In the remaining six states, the GOP had majorities in both houses of the legislature, and often a "supermajority" of at least two-thirds control in both houses. A party that represents conservative Christians should have been able to use its large majorities to pass the anti-evolution bills easily.

Furthermore, the reason for the anti-evolution movement's lack of success cannot be that Republicans have seen the result of court decisions and decided not to waste time on legislation that is certain to be invalidated by federal judges. The right to have an abortion has long been protected by judges, and yet Republican state legislators never tire of passing laws to try to suppress a woman's right to choose. So why do socially conservative legislatures keep failing to pass anti-evolution laws, when they keep passing anti-abortion laws? Perhaps the answer has something to do with politics.

## Lost in Multidimensional Space

Political scientists like to point out that parties are composed of factions, or "coalitions of enemies," combining interests and ideologies from different dimensions of the political spectrum into an uneasy alliance.[70] Perhaps, however, that is an exaggeration. Maybe a party imposes an association of many different people and social forces that are mostly indifferent to one another but can be aroused to quarrel under certain conditions. In either case, the elected officials and the voters, activists, and investors who form the factions find themselves cooperating most of the time, but antagonistic upon occasion. Sometimes, party cohorts that disagree strongly on one issue engage in campaigns of sabotage against each other in private while they are singing hymns of cooperation in public. If enough inner conflict lasts long enough, it will split the party, as large chunks of it go elsewhere. But if the conflict can be managed, the party will stay together in general, while party leaders are busy putting down rebellions by specific groups over certain issues. As political scientist Elaine Kamarck summarizes the rule, "the scholar or practitioner of change needs to be aware of the fact that, on any individual issue, there's a high probability that individual factions end up being more important than party."[71]

In the Reaganite Republican party, the alliance between the economic conservative faction—that is, those people who are attempting to further the interests of business—and the social conservative faction has never been a completely happy one. As Miller and Schofield observe, "If anyone had argued prior to 1960 that Wall Street and populists would happily join hands within the same political party, both sides would have laughed at the idea. . . . Today the question has become, Who controls the Republican Party—social conservatives with their beliefs in moral interventionism, or the proponents of business?"[72] The answer would seem to be that different factions control the party at different times and places. Moreover, sometimes the business wing of the party, which influences representatives by being a past and potentially future source of campaign contributions, realizes that it will be outvoted by the social conservative wing of the party, which looks to activists and ordinary citizens as the sources of its power. When business-oriented Republicans feel that business values or business interests are threatened, they will sometimes quietly betray the coalition. The betrayal can be successful if the saboteurs are able to find allies outside the party.

In the case of evolution, there is a manifest conflict between the pro-business (often termed "The Establishment") and social conservative (often termed the

"Christian Right") wings of the Republican party. Business leaders know that in the broad sweep of a likely future, their prosperity depends upon having access to an educated workforce.[73] If young Americans arrive in the labor market unable to think scientifically and contemptuous of scientific values, then American business will be in trouble. Some businesses might be able to survive by hiring stupid and ignorant workers for tasks that only require those qualifications. Other businesses might prosper by hiring educated, scientifically oriented immigrants from countries that do not suffer from fundamentalist anti-intellectualism. But most Establishment Republicans know that in the long run they cannot afford to rely on a workforce that has been miseducated and misfitted for a modern economy.[74]

And so, on this issue, they find allies outside their party. The allies consist of two types. Democrats, in general, feel no ambivalence about evolution. Although they are aware that the American public, in a diffuse sort of way, would like to see creationism taught alongside evolution, their activists and investors are solidly pro-science. And Democratic officeholders, having no fundamentalist constituencies to placate, are not inclined to take an anti-science stand in a legislature. Meanwhile, various technical elites, including scientists and public school teachers, are overwhelmingly in favor of keeping science education scientific.[75] There is at least anecdotal evidence to suggest that the business community has frequently worked with educational groups to ensure that state science standards reflect the opinions of scientists rather than those of the mass public.[76]

Thus the apparently anomalous trend toward anti-evolution bills failing in state legislatures dominated by Republicans. When such bills are introduced by social conservatives, the economic conservatives cooperate with Democrats to kill them. Usually, the bills are euthanized in committees, where it is easy for Establishment Republicans to smother the legislation away from any publicity. But even when a bill makes it out of committee, as in the Oklahoma legislature in 2013, the socially conservative wing of the party is not able to overcome the temporary coalition of some of its allies with its enemies. When, in the unusual circumstance that an anti-evolution bill is passed and signed by a governor, it is certain to be quashed by a federal court.

In Chapter Five, I posed the question of why Judge John E. Jones, who presided over the *Kitzmiller* trial in Dover, rendered such a devastatingly anti–Intelligent Design decision. Jones, a conservative Republican appointed to the federal bench by George W. Bush, might have been expected to come down on the side of creationism in the public schools, or at least to duck the big issue with a technical holding. Instead, his decision was so uncompromising that even ID proponents were forced to concede, "It is hard to imagine a court decision that could have been formulated more negatively against intelligent design."[77] Perhaps at some time in the future, an historian will write a book explaining who assigned Jones to the case, and why, and explicating Jones' judicial ideology. In the meantime, I will speculate that Jones is exclusively a staunch economic conservative, a business-oriented jurist who fears the advancement of Christian creationism into the public schools as much as any Democrat, and that whoever assigned him to preside over the Dover case did so for precisely that reason.

Within the Republican party, the contrast between the treatment of evolution and the treatment of abortion is instructive. Abortion, like evolution, is one of the rallying issues of the Christian Right. But whether abortion is or is not prohibited in an individual state, or in the country as a whole, is largely irrelevant to the future of business. The presence of abortion might conceivably be relevant to the number of workers in a labor force—more abortions, fewer babies, and therefore fewer adults—but it does not impinge on the quality of a workforce. Whether abortion stays available or not will have no impact on American workers' scientific competence. Business leaders no doubt have personal opinions about the issue of abortion, but they do not have opinions on that subject *as business leaders*. Therefore, when anti-abortion bills are introduced into state legislatures, Establishment conservatives are willing to maintain the coalition and vote with members of the Christian Right. When anti-evolution bills are introduced, however, many members of the Republican coalition go over to the enemy.

Thus, when we look at the actual functioning of American politics in the present era, as opposed to public opinion polls and party platforms, the future of science education looks somewhat more secure. A coalition of minorities—business, scientists, educators, pro-science opinion leaders, judges—has generally been able to stifle the implementation of the people's will. Given the alignment of forces, their access to campaign finances, and their position within an independent federal judiciary, they will probably be able to maintain elite dominance of public education for the foreseeable future. And good for them.

## Back to the Future of Democracy

This conclusion, however, brings us back to the problem of democracy. Although it is often discussed as though it were a social contraption consisting of elections and government institutions, democracy is at its heart a moral theory. It addresses the question of what sort of relations of power can be legitimate—morally grounded—and, therefore, when citizens have a responsibility to obey a law with which they might not agree and when they would be justified in rebelling.

Locke and Rousseau, the classic democratic philosophers in the seventeenth and eighteenth centuries, not being able to observe functioning democracies in their own eras, had to consider this problem of legitimacy in the abstract.[78] Their argument was that only the people's participation—"consent of the governed," in its Lockean phrasing—could infuse governmental actions with legitimacy, and therefore compel obedience from the citizen on the street. As a practical matter, they recommended majority rule as the best measure of the people's will.

Jefferson famously borrowed most of Locke's argument for the "Declaration of Independence," which was mainly a philosophical complaint that the British government, by refusing to grant its American colonies the right to participate in its parliament, had voided its moral claim to exact legitimate obedience. When Hamilton, Madison, and the lesser authors of the Constitution launched a campaign to sell it to the public during the ratification campaign of 1787, they also pointed to the people's participation as the "fountain" from which legitimacy flowed.[79] But they, having observed democratic governments in operation, were also greatly concerned about the potential for majoritarian misbehavior. The people might be sovereign, but the people must not be allowed to trample the rights of the minority, especially property holders. Hamilton's solution to the problem of democratic overreach, as I discussed in Chapter Five, was an independent judiciary. Madison's twin solutions were a republic so large that it would prevent a majority faction from coalescing and a set of competing institutions that would prevent each other from abusing minority rights.[80]

As the republic advanced through the centuries, astute observers tried to come to terms with flaws in its democratic functioning. Various thinkers have tried to address the problems that have arisen because some people are more organized or intense than others, and therefore exert more influence over practical politics.[81] By far the most thought and energy, however, has gone into the difficulties that a theory basing legitimacy on the people's participation encounters when faced with the monstrous inequalities of political influence created by inequalities in the economic system.[82] Some theorists have refined the problem to one of unequal access to the media in particular, rather than unequal access to economic resources in general.[83]

Nevertheless, however they define the problem to be solved, the overwhelming majority of writers on the topic of democratic imperfections have addressed the issue from a stance based on the assumption that the sticking point is that the people's preferences are not being implemented. In almost every commentary on democracy—and libraries of the stuff are published each year—the author observes that the legitimacy of the system is in danger because various social forces and institutions are preventing mass opinions from governing the country. Further, the basic pro-democratic discourse has sprouted subdiscourses, so that advocates of "cosmopolitan," "agonistic," "republican," "monitory," "strong," "radical," and "associative" democracy contrast their ideals with the deficient reality.[84]

Mainly, observers on the left of the political spectrum make this complaint. Even the American Political Science Association has gotten into the act, expressing alarm about the consequences for democratic legitimacy of the ongoing trend toward economic inequality.[85] But there are also those on the right of the spectrum lamenting that the United States is undergoing "an erosion of moral adherence to this political system" because the "citizens of this democratic republic are deemed to lack the competence for self-government."[86] Left or right, it seems that everybody worries that the people do not have enough influence in the world's greatest democracy.

But underneath, and rarely expressed clearly, there is the worry that, in regard to some issues, the people are a bunch of damn fools, and therefore cannot be permitted to chart the future of the country. An elite of "guardians," whether of the aristocracy or The Party, must govern. This attitude is a tradition that goes back to Plato, of course, but it is not a democratic tradition, and it has no partisans willing to publish their thoughts in the modern United States. Hamilton's independent judiciary is a partial exception, but it was intended to block temporary surges of popular zealotry in the short run. Its power was never supposed to cancel the basic right of the public to set long-run government policy.

Thus, when Robert Dahl surveyed social thought for spokespeople for the "guardianship" theory of government for a 1989 book, he could discover no systematic defenses of that nondemocratic theory except in B.F. Skinner's 1948 novel *Walden Two*.[87] Skinner, a psychologist, thought that psychologists should run the country, using operant conditioning to form human character into a citizenry of Platonic perfection who would need no guardians. He had no theory of legitimacy, and the book only qualifies as political philosophy under the most relaxed definition. In other words, although the people may sometimes be unworthy of self-government, at the largest and most inclusive level all American philosophy agrees that they should nevertheless have the final say over government policy.

At the small and specific level, however, many thinkers are willing to forget their democratic commitments and advocate rule by guardians. Especially in the area of public school curriculum, the people are held to be incompetent to decide the content of the lessons presented to their own children. As journalist Walter Lippmann, contemplating with dismay the spectacle of the citizens of Tennessee's preferences on display in the Scopes trial, made the case in 1927:

> The votes of the majority [should] have no intrinsic bearing on the conduct of a school. . . . Guidance for a school can only come from educators, and the question of what shall be taught as biology can be determined only by biologists. The votes of a majority do not settle anything here and they are entitled to no respect whatsoever.[88]

Although they rarely state the principle so forthrightly, the many scientists, educators, and social commentators who have written truckloads of books and articles exposing the frauds of, first, scientific creationism and lately, intelligent design, entirely agree with Lippmann. Education, they imply, is too important to be left to the people's judgment. Education will mold the future, the world of the future will demand scientific understanding, and we cannot permit future citizens to carry around a bunch of reactionary superstitions in their heads.

Therefore, we liberal educators, and we non-partisan scientists, will enter into a tacit coalition with economically conservative business leaders, using the First Amendment, our access to the media, our time, and their money, to make sure that the foolish majority who are supposed to determine government policy do not actually do so. And we will do it in the name of democracy.

I must therefore end this book as I began it, on a note of uneasy ambivalence. On the facts of the case, in the contest between the advocates of acknowledging the truth of the theory of natural selection, in schools and elsewhere, versus the advocates of ignoring evolution at the least, or teaching creationism at the most, the winner is unambiguous. Evolution is true; creationism, however labeled or packaged, is false. But as a matter of democratic legitimacy, the spectacle of a tacit coalition of elites, suppressing the desires of generations of parents to instruct their children according to their own beliefs and values, is profoundly unsettling.

As I argued in Chapter Four, there is a faint ray of hope that an appropriate education strategy, properly implemented, might over time change public opinion so that pro-evolution attitudes become the majority position, and the problem for democratic theory disappears. But the evidence that such a wholesome change might come about is so diaphanous that it can inspire only the weakest optimism.

And so, on this last page as on the first, I see a serious Problem, but I am not confident that a Solution can be found.

## Notes

1 This letter is quoted and discussed in Adair, Douglass, "Was Alexander Hamilton a Christian Statesman?" in *Fame and the Founding Fathers: Essays by Douglass Adair* (Indianapolis: Liberty Fund, 1974), 222–223; the original essay, co-authored with Marvin Harvey, appeared in *William and Mary Quarterly*, 3rd Ser., XII (1955), 308–329.
2 Ibid.
3 Adorno, Theodor W., Else Frenkel-Brunswik, Daniel J. Levinson, and R. Nevitt Sanford, *The Authoritarian Personality* (New York: Harper and Row, 1950).
4 McClosky, Herbert, "Conservatism and Personality," in Nelson W. Polsby, Robert A. Dentler, and Paul A. Smith, eds., *Politics and Social Life: An Introduction to Political Behavior* (Boston: Houghton Mifflin, 1963), 226; originally published in *The American Political Science Review*, 52 (March 1958), 27–45.
5 Jost, John T., Jack Glaser, Arie W. Kruglanski, and Frank J. Sullaway, "Political Conservatism as Motivated Social Cognition," *Psychological Bulletin*, 129, no. 3 (2003), 339–375.
6 Beginning with Christie, Richard and Marie Jahoda, eds., *Studies in the Scope and Method of the Authoritarian Personality* (Glencoe, Ill.: Free Press, 1954).
7 Charney, Evan, "Genes and Ideologies," *Perspectives on Politics*, 6, no. 2 (June 2008), 308–309.
8 Rokeach, Milton, *The Open and Closed Mind: Investigations into the Nature of Belief Systems and Personality Systems* (New York: Basic Books, 1960), 117, 121–123, 226.
9 Ibid., 115–116.

10  Ibid., 55–56.
11  Ibid., 111, 351.
12  Sherkat, Darren E. and Christopher G. Ellison, "The Cognitive Structure of a Moral Crusade: Conservative Protestantism and Opposition to Pornography," *Social Forces*, 75, no. 3 (March 1997), 961, 963, 965, 968.
13  Burge, Ryan, "Using Matching to Investigate the Relationship between Religion and Tolerance," *Politics and Religion*, 6, no. 2 (June 2013), 264–281; Putnam, Robert D. and David E. Campbell, *American Grace: How Religion Divides and Unites Us* (New York: Simon and Schuster, 2010), 479–482; Putnam and Campbell use a more general measure of "religiosity" than a specific measure of fundamentalism, but general research in this area makes it very clear that fundamentalists are intensely religious; see also Ellison, Christopher G. and Marc A. Musick, "Southern Intolerance: A Fundamentalist Effect?" *Social Forces*, 72, no. 2 (December 1993), 382–384.
14  Evans, E. Margaret, "Cognitive and Contextual Factors in the Emergence of Diverse Belief Systems: Creation versus Evolution," *Cognitive Psychology*, 42 (2001), 229 and passim; Freeman, Patricia K. and David J. Houston, "The Biology Battle: Public Opinion and the Origins of Life," *Politics and Religion*, 2, no. 1 (April 2009), 61–68.
15  Putnam and Campbell, *American Grace*, op. cit., 27, 272.
16  Ellison and Musick, "Southern Intolerance," op. cit., 389.
17  Conger, Kimberly H. and Bryan T. McGraw, "Religious Conservatives and the Requirements of Citizenship: Political Autonomy," *Perspectives on Politics*, 6, no. 2 (June 2008), 253–266.
18  McDaniel, Eric L. and Christopher Ellison, "God's Party? Race, Religion, and Partisanship over Time," *Political Research Quarterly*, 61, no. 2 (June 2008), 180–191.
19  Sherkat and Ellison, "Cognitive Structure," op. cit., 972.
20  D'Antonio, William V., Steven A Tuch, and Josiah R. Baker, *Religion, Politics, and Polarization: How Religiopolitical Conflict Is Changing Congress and American Democracy* (Lanham, Md.: Rowman and Littlefield, 2013), 2, 8, 89; Freeman and Houston, "The Biology Battle," op. cit., 59–63.
21  The classic exposition of this theme is by Downs, Anthony, *An Economic Theory of Democracy* (New York: Harper and Row, 1957).
22  Schofield, Norman, "Instability of Simple Dynamic Games," *Review of Economic Studies*, 45, no. 3 (1978), 575–594; McKelvey, Richard, "General Conditions for Global Intransitivities in Formal Voting Models," *Econometrica*, 47 (1979), 1085–1111; Schwartz, Thomas, *The Logic of Collective Choice* (New York: Columbia University Press, 1986).
23  Ferguson, Thomas and Joel Rogers, *Right Turn: The Decline of the Democrats and the Future of American Politics* (New York: Hill and Wang, 1986), 45.
24  Miller, Gary and Norman Schofield, "Activists and Partisan Realignment in the United States," *American Political Science Review*, 97, no. 2 (May 2003), 245–260. I should note here that Miller and Schofield provide a theoretical model of only two groups, candidates and activists, whereas my discussion involves three groups. I have simply extended the basic principles of their analysis to investors, who can be considered another type of activist. To Miller and Schofield, an activist is someone who invests time. I am adding those who invest significant amounts of money.
25  Workman, Samuel, Bryan D. Jones, and Ashley E. Jochim, "Information Processing and Policy Dynamics," *Policy Studies Journal*, 37, no. 1 (2009), 75–92.

26 Lipset, Seymour Martin, *Political Man: The Social Basis of Politics* (Garden City, N.Y.: Doubleday, 1963), 230; originally published 1960.
27 Wald, Kenneth D. and Clyde Wilcox, "Getting Religion: Has Political Science Rediscovered the Faith Factor?" *American Political Science Review*, 100, no. 4 (November 2006), 523–529; Kettell, Steven, "Has Political Science Ignored Religion?" *PS: Political Science and Politics*, 45, no. 1 (January 2012), 93–100.
28 Key, V.O., Jr., *Southern Politics* (New York: Random House, 1949), 5.
29 Ibid., 307.
30 Ibid., 267.
31 Reed, John Shelton, *The Enduring South: Subcultural Persistence in Mass Society* (Chapel Hill: University of North Carolina Press, 1974), 57.
32 Prindle, David F., *The Paradox of Democratic Capitalism: Politics and Economics in American Thought* (Baltimore: Johns Hopkins University Pres, 2006), 50–53, 58, 65, 71–73, 136, 137.
33 Hofstadter, Richard, *The American Political Tradition and the Men Who Made It* (New York: Random House, 1948).
34 Ibid., 125–138.
35 Kleppner, Paul, *The Cross of Culture: A Social Analysis of Midwestern Politics, 1850–1900* (New York: Free Press, 1970), 35–91; Jensen, Richard, *The Winning of the Midwest: Social and Political Conflict, 1888–96* (Chicago: University of Chicago Press, 1971), 58–88.
36 Burnham, Walter Dean, *Critical Elections and the Mainsprings of American Politics* (New York: W.W. Norton, 1970), 71–90.
37 Miller and Schofield, "Activists and Partisan Realignment," op. cit., 249.
38 Layton, Azza Salama, *International Politics and Civil Rights Policies in the United States, 1941–1960* (Cambridge: Cambridge University Press, 2000).
39 Jarboe, Jan, "Lady Bird Looks Back," *Texas Monthly*, December 1994, 117.
40 Burnham, *Critical Elections*, 118–119; Charles S. Bullock III, Donna R. Hoffman, and Ronald Keith Gaddie, "The Consolidation of the White Southern Congressional Vote," *Political Research Quarterly*, 58, no. 2 (June 2005), 231–244.
41 Martin, William, *With God on Our Side: The Rise of the Religious Right in America* (New York: Broadway Books, 1996), 213.
42 Ibid.
43 Ibid., 216–217; my characterization of the John Anderson campaign is based on my own experience as an adviser to that campaign's spokesperson in Texas.
44 Ibid., 217–218.
45 Ferguson and Rogers, *Right Turn*, 34–35.
46 Miller and Schofield, "Activists and Partisan Realignment," op. cit., 246–247.
47 Shogan, Colleen J., "Anti-intellectualism in the Modern Presidency: A Republican Populism," *Perspectives on Politics*, 5, no. 2 (June 2007), 300.
48 Bush quoted in Forrest, Barbara and Paul R. Gross, *Creationism's Trojan Horse: The Wedge of Intelligent Design* (Oxford: Oxford University Press, 2004), 254.
49 Salisbury, Robert H., "An Exchange Theory of Interest Groups," *Midwest Journal of Political Science*, 13 (Spring 1969), 1–32; Casamayou, Maureen, "Collective Entrepreneurialism and Breast Cancer Advocacy," in Allan J. Cigler and Burdett A. Loomis, eds., *Interest Group Politics*, 6th ed. (Washington, D.C.: CQ Press, 2002), 79–94; Leech, Beth L., Frank R. Baumgartner, Jeffrey M. Berry, Marie Hojnacki, and David C. Kimball, "Organized Interests and Issue Definition in Policy Debates," in ibid., 275–292.

50 Information about Johnson from Forrest and Gross, *Creationism's Trojan Horse*, op. cit., 16-17.
51 Dawkins, Richard, *The Blind Watchmaker: Why the Evidence of Evolution Reveals a Universe without Design*, 2nd ed. (New York: W.W. Norton, 1996), first published 1986; Denton, Michael, *Evolution: A Theory in Crisis* (Chevy Chase, Md.: Adler and Adler, 1986).
52 Johnson quoted in Forrest and Gross, *Creationism's Trojan Horse*, op. cit., 17.
53 Johnson, Phillip E., *Darwin on Trial*, 2nd ed. (Downers Grove, Ill.: InterVarsity Press, 1993); first edition published 1991.
54 Ibid., 14.
55 A discussion of the meeting at Pajaro Dunes occupies the first section of the documentary *Unlocking the Mystery of Life*, which is both an introduction to and a sales pitch for the doctrine of Intelligent Design; it was produced in 2002 by Illustra Media (www.ilustramedia.com) at the behest of the Discovery Institute (www.Discovery.org); Forrest and Gross, *Creationism's Trojan Horse*, op. cit., 265–266; Dembski, William, "Preface," in William Dembski, ed., *Darwin's Nemesis: Phillip Johnson and the Intelligent Design Movement* (Downers Grove, Ill.: InterVarsity Press, 2006), 14.
56 Dembski, *Darwin's Nemesis*, op. cit., 12–13.
57 Forrest and Gross, *Creationism's Trojan Horse*, op. cit., 22–23.
58 Micklethwait, John and Adrian Wooldridge, *The Right Nation: Conservative Power in America* (New York: Penguin, 2004), 158–159.
59 Dembski, William A., "Dealing with the Backlash against Intelligent Design," in Dembski, ed., *Darwin's Nemesis*, op. cit., 86–87.
60 Forrest and Gross, *Creationism's Trojan Horse*, op. cit., 206, 223; Plantinga, Alvin, "A Modest Proposal," in Robert T. Pennock, ed., *Intelligent Design Creationism and Its Critics: Philosophical, Theological, and Scientific Perspectives* (Cambridge, Mass.: MIT Press, 2001), 790.
61 Forrest and Gross, *Creationism's Trojan Horse*, op. cit., 204, 223.
62 Frank, Thomas, *What's the Matter with Kansas? How Conservatives Won the Heart of America* (New York: Henry Holt and Company, 2004), 210.
63 Kitcher, Philip, "Born-Again Creationism," in Pennock, *Intelligent Design Creationism*, op. cit., 263.
64 Forrest and Gross, *Creationism's Trojan Horse*, op. cit., 221–222; Frank, *What's the Matter with Kansas?*, op. cit., 205–208.
65 Santorum Amendment quoted in Forrest and Gross, *Creationism's Trojan Horse*, op. cit., 240–241.
66 Ibid., 243.
67 Blake, Mariah, "Revisionaries: How a Group of Conservatives Is Rewriting Your Kids," *Washington Monthly*, website accessed July 2, 2013; for an example of a McLeroy rant on YouTube, go to the site and insert "Don McLeroy on Stephen Jay Gould" into the search engine.
68 "Texas State Board of Education Primary Delivers Upset," *Daily Kos* website, March 2, 2010; editorial, "Elect the Capable, Shun Ideologues," *Austin American-Statesman*, October 24, 2012, A10.
69 Information about potential legislation, and its demise, from the website of the National Center for Science Education, http://ncse.com/; information about party balance from state legislative websites.
70 Miller and Schofield, "Activists and Partisan Realignment," op. cit., 249.

71 Kamarck, Elaine C., *How Change Happens—or Doesn't: The Politics of U.S. Public Policy* (Boulder, Colo.: Lynne Rienner, 2013), 74.
72 Miller, Gary and Norman Schofield, "The Transformation of the Republican and Democratic Party Coalitions in the U.S.," *Perspectives on Politics*, 6, no. 3 (September 2008), 439.
73 Wolbrecht, Christina and Michael.T. Hartney, "'Ideas about Interests': Explaining the Changing Partisan Politics of Education," *Perspectives on Politics*, 12, no. 3 (September 2014), 616, 618.
74 Other than in some passages in Miller and Schofield, "The Transformation," I have not been able to find empirical scholarship addressed to the topic of the split within the Republican party. But the conflict between economic conservatives and social conservatives is a common subject for journalists and bloggers. For example: Patterson, Robert W., "Fiscal Conservatism Is Not Enough: What Social Conservatives Offer the Party of Lincoln," *The Family in America* (Spring 2010), 120, 121; Blodget, Henry, "It Is Infuriating that I Can't Vote for a Fiscal Conservative without also Supporting Religious Aggressives," *Business Insider*, August 23, 2012; Krugman, Paul, "The Fix Isn't In: Eric Cantor and the Death of a Movement," *New York Times*, (June 12, 2014).
75 Berkman, Michael and Eric Plutzer, *Evolution, Creationism, and the Battle to Control America's Classrooms* (Cambridge: Cambridge University Press, 2010), 125.
76 Ibid., 223.
77 Dembski, "Preface" to *Darwin's Nemesis*, op. cit., 19.
78 Locke, John, *Second Treatise of Government* (Indianapolis: Hackett, 1980), originally published 1690; Rousseau, Jean Jacques, *The Social Contract* (New York: Penguin, 1968).
79 Hamilton, Alexander, John Jay, and James Madison, *The Federalist* (New York: Modern Library, 1937), 141, 227, 327.
80 Ibid., *#10* and *#51*, 53–61, 335–340.
81 McConnell, Grant, *Private Power and American Democracy* (New York: Alfred A. Knopf, 1966); Dahl, Robert A., *A Preface to Democratic Theory* (Chicago: University of Chicago Press, 1956); Olson, Mancur, *The Logic of Collective Action: Public Goods and the Theory of Groups* (Cambridge, Mass.: Harvard University Press, 1965).
82 This literature is too vast to cover in detail; as a small sample, see Cohen, Joshua and Joel Rogers, *On Democracy: Toward a Transformation of American Society* (New York: Penguin Books, 1983); Prindle, David F., *The Paradox of Democratic Capitalism: Politics and Economics in American Thought* (Baltimore: Johns Hopkins University Press, 2006).
83 McChesney, Robert W., *Rich Media, Poor Democracy: Communication Politics in Dubious Times* (Champaign: University of Illinois Press, 1999).
84 Pateman, Carol, "Participatory Democracy Revisited," *Perspectives on Politics*, 10, no. 1 (March 2012), 7; Prindle, *Paradox of Democratic Capitalism*, op. cit., 256.
85 American Political Science Association Task Force on Inequality and American Democracy, "APSA Task Force in an Age of Rising Inequality," *Perspectives on Politics*, 2, no. 4 (December 2004), 651–666.
86 The editors of *First Things*, "Introduction" to *The End of Democracy*, Mitchell S. Muncy, ed. (Dallas: Spence Publishing Company, 1997), 6–7.
87 Dahl, Robert A., *Democracy and Its Critics* (New Haven, Conn.: Yale University Press, 1989), 52–53; Skinner, Burrhus F., *Walden Two* (New York: Macmillan, 1948).
88 Lippmann quoted in Berkman and Plutzer, *Evolution, Creationism, and the Battle to Control America's Classrooms*, op. cit., 11.

# BIBLIOGRAPHY

Adair, Douglass, "Was Alexander Hamilton a Christian Statesman?" in *Fame and the Founding Fathers: Essays by Douglass Adair* (Indianapolis: Liberty Fund, 1974), 222–223.
Adorno, Theodor W., Else Frenkel-Brunswik, Daniel J. Levinson, and R. Nevitt Sanford, *The Authoritarian Personality* (New York: Harper and Row, 1950).
Adrian, Desmond and James Moore, *Darwin: The Life of a Tormented Evolutionist* (New York: W.W. Norton, 1991).
Alcock, John, *The Triumph of Sociobiology* (Oxford: Oxford University Press, 2001).
Alroy, John, "Constant Extinction, Constrained Diversification, and Uncoordinated Stasis in North American Mammals," *Palaeo*, 127 (1996), 285–311.
Alters, Brian, "Evolution in the Classroom," in Eugenie C. Scott and Glenn Branch, eds., *Not in Our Classrooms: Why Intelligent Design Is Wrong for Our Schools* (Boston: Beacon Press, 2006), 105–129.
American Political Science Association Task Force on Inequality and American Democracy, "APSA Task Force in an Age of Rising Inequality," *Perspectives on Politics*, 2, no. 4 (December 2004), 651–666.
Anderson, Mike L., "The Effect of Evolutionary Teaching on Students' Views of God as Creator," *Journal of Theology for Southern Africa*, 87 (1994), 69–73.
Angier, Natalie, *The Canon: A Whirligig Tour of the Beautiful Basics of Science* (Boston: Houghton Mifflin, 2007).
Aquinas, Thomas, "Of Usury," in Philip C. Newman, Arthur D. Gayer, and Milton H. Spencer, eds., *Sources in Economic Thought* (New York: W.W. Norton, 1954), 19–21 (this is a selection from *Summa Theologica*).
Armstrong, Karen, *The Battle for God: A History of Fundamentalism* (New York: Random House, 2000).
"A Special Report," *Christian Harvest Times*, (June 1980), 1.
Atran, Scott, *In Gods We Trust: The Evolutionary Landscape of Religion* (Oxford: Oxford University Press, 2002).

Bacon, Francis, *Magna Instauratio* (originally published 1620), in Richard Foster Jones, ed., *Francis Bacon: Essays, Advancement of Learning, New Atlantis, and Other Pieces* (New York: Odyssey Press, 1937), 308, 310.

———, *The Advancement of Learning* (originally published 1605) in Richard Foster Jones, ed., *Francis Bacon: Essays, Advancements of Learning, New Atlantis, and Other Pieces* (New York: Odyssey Press, 1937), 214.

Barash, David, *The Whisperings Within: Evolution and the Origin of Human Nature* (New York: Penguin, 1979), 90.

Barham, James, "Why I Am Not a Darwinist," in William A. Dembski, ed., *Uncommon Dissent: Intellectuals Who Find Darwinism Unconvincing* (Wilmington, Del.: ISI Books, 2004), 177–191.

Barnes, Barry, David Bloor, and John Henry, *Scientific Knowledge: A Sociological Analysis* (Chicago: University of Chicago Press, 1996).

Baugh, Carl, *Why Do Men Believe Evolution against All Odds?* (Bethany, Okla.: Bible Belt Publishing, 1999).

Behe, Michael J., *The Edge of Evolution: The Search for the Limits of Darwinism* (New York: Free Press, 2007).

———, "Molecular Machines: Experimental Support for the Design Inference," in Robert T. Pennock, ed., *Intelligent Design Creationism and Its Critics: Philosophical, Theological, and Scientific Perspectives* (Cambridge, Mass.: MIT Press, 2001), 241–256.

———, *Darwin's Black Box: The Biochemical Challenge to Evolution* (New York: Simon and Schuster, 1996).

Berkman, Michael and Eric Plutzer, *Evolution, Creationism, and the Battle to Control America's Classrooms* (New York: Cambridge University Press, 2010), 35–39.

Berman, Marshall, *All that Is Solid Melts into Air: The Experience of Modernity* (New York: Simon and Schuster, 1982).

Bishop, George F., Randall K. Thomas, and Jason A. Wood, "Measurement Error, Anomalies, and Complexities in Americans' Beliefs about Human Evolution," *Survey Practice*, 3, no. 1 (2010). Available at: www.surveypractice.org: accessed June 25, 2014.

Bishop, George F., Randall K. Thomas, Jason A. Wood, and Misook Gwon, "Americans' Scientific Knowledge and Beliefs about Human Evolution in the Year of Darwin," *Reports of the National Center for Science Education*, 30, no. 3 (May–June 2010), 16–18.

Blake, Mariah, "Revisionaries: How a Group of Conservatives Is Rewriting Your Kids' Textbooks," *Washington Monthly*. Available at: www.washingtonmonthly.com/features/2010/1001.blake.html: accessed June 28, 2014.

Blanchard, Troy C., John P. Bartkowski, Todd L. Matthews, and Kent R. Kerley, "Faith, Morality and Mortality: The Ecological Impact of Religion on Population Health," *Social Forces*, 86, no. 4 (June 2008).

Bleier, Ruth, *Science and Gender: A Critique of Biology and Its Theories on Women* (Oxford: Pergamon Press, 1984), vii.

Blodget, Henry, "It Is Infuriating that I Can't Vote for a Fiscal Conservative without Also Supporting Religious Aggressives," *Business Insider* (August 23, 2012).

Bork, Robert H., *The Tempting of America: The Political Seduction of the Law* (New York: Simon and Schuster, 1990), 7.

Bowler, Peter J., *Evolution: The History of an Idea*, 3rd ed. (Berkeley: University of California Press, 2003).

Boyd, Richard, "Metaphor and Theory Change," in Andrew Ortory, ed., *Metaphor and Thought* (London: Cambridge University Press, 1979).
Boyer, Pascal, *Religion Explained: The Evolutionary Origins of Religious Thought* (New York: Basic Books, 2001).
Bradley, Walter L., "Phillip Johnson and the Intelligent Design Movement," in William A. Dembski, ed., *Darwin's Nemesis: Phillip Johnson and the Intelligent Design Movement* (Leicester, England: Inter-Varsity Press, 2006), 305–314.
Breyer, Stephen, *Active Liberty: Interpreting Our Democratic Constitution* (New York: Random House, 2005), 99.
Brooke, John Hedley, *Science and Religion: Some Historical Perspectives* (Cambridge: Cambridge University Press, 1991).
Brooks, Cleanth and Robert Penn Warren, *Understanding Poetry* (New York: Holt, Rinehart, and Winston, 1938).
Brown, Mark B., *Science in Democracy: Expertise, Institutions, and Representation* (Cambridge, Mass.: MIT Press, 2009), 198.
Brown, Richard Harvey, "Rhetoric and the Science of History: The Debate between Evolutionism and Empiricism as a Conflict of Metaphors," *Quarterly Journal of Speech*, 72 (1986), 148–161.
Brown, Walt, *In the Beginning: Compelling Evidence for Creation and the Flood*, 8th ed. (Phoenix, Ariz.: Center for Scientific Creation, 2008).
Buller, David J., *Adapting Minds: Evolutionary Psychology and the Persistent Quest for Human Nature* (Cambridge, Mass.: MIT Press, 2005).
Bullock, Charles S. III, Donna R. Hoffman, and Ronald Keith Gaddie, "The Consolidation of the White Southern Congressional Vote," *Political Research Quarterly*, 58, no. 2 (June 2005), 231–244.
Burge, Ryan, "Using Matching to Investigate the Relationship between Religion and Tolerance," *Politics and Religion*, 6, no. 2 (June 2013), 264–281.
Burnham, Walter Dean, *Critical Elections and the Mainsprings of American Politics* (New York: W.W. Norton, 1970).
Buss, David M., *Evolutionary Psychology: The New Science of the Mind* (Boston: Pearson, 2004).
———, *The Evolution of Desire: Strategies of Human Mating* (New York: HarperCollins, 1994).
Carnegie, Andrew, "Wealth," in Michael B. Levy, ed., *Political Thought in America: An Anthology*, 2nd ed., selected readings (Chicago: The Dorsey Press, 1988), 331–336.
Carruthers, Peter and Andrew Chamberlain, "Introduction," in Carruthers and Chamberlain, eds., *Evolution and the Human Mind: Modularity, Language, and Meta-Cognition* (Cambridge: Cambridge University Press, 2000).
Casamayou, Maureen, "Collective Entrepreneurialism and Breast Cancer Advocacy," in Allan J. Cigler and Burdett A. Loomis, eds., *Interest Group Politics*, 6th ed. (Washington, D.C.: CQ Press, 2002), 79–94.
Charney, Davida, "A Study in Rhetorical Reading: How Evolutionists Read 'The Spandrels of San Marco,'" in Jack Selzer, ed., *Understanding Scientific Prose* (Madison: University of Wisconsin Press, 1993).
Charney, Evan, "Genes and Ideologies," *Perspectives on Politics*, 6, no. 2 (June 2008), 308–309.
Chinsamy, Anusuya and Eva Planany, "Accepting Evolution," *Evolution*, 62, no. 1 (2007), 248–254.

Chong, Dennis, "Creating Common Frames of Reference on Political Issues," in Diana C. Mutz, Paul M. Sniderman, and Richard A. Brody, eds., *Political Persuasion and Attitude Change* (Ann Arbor: University of Michigan Press, 1996).
Christie, Richard and Marie Jahoda, eds., *Studies in the Scope and Method of the Authoritarian Personality* (Glencoe, Ill.: Free Press, 1954).
Cohen, H. Floris, *The Scientific Revolution: A Historiographical Inquiry* (Chicago: University of Chicago Press, 1994).
Cohen, Joshua and Joel Rogers, *On Democracy: Toward a Transformation of American Society* (New York: Penguin Books, 1983).
Collins, Francis S., *The Language of God: A Scientist Presents Evidence for Belief* (New York: Free Press, 2006).
Condit, Celeste M., Benjamin R. Bates, Ryan Galloway, Sonja Brown Givens, Caroline K. Haynie, John W. Jordan, Gordon Stables, and Hollis Marshall West, "Recipes or Blueprints for Our Genes? How Contexts Selectively Activate the Multiple Meanings of Metaphors," *Quarterly Journal of Speech*, 88, no. 3 (August 2002), 303–325.
Conger, Kimberly H. and Bryan T. McGraw, "Religious Conservatives and the Requirements of Citizenship: Political Autonomy," *Perspectives on Politics*, 6, no. 2 (June 2008), 253–266.
Cornfield, Richard, *Architects of Eternity: The New Science of Fossils* (London: Headline Book Publishing, 2001).
Cosmides, Leda and John Tooby, "The Psychological Foundations of Culture," in Jerome H. Barkow, Leda Cosmides, and John Tooby, eds., *The Adapted Mind: Evolutionary Psychology and the Generation of Culture* (New York: Oxford University Press, 1992), 19–136.
Cosmides, Leda, John Tooby, and Jerome Barkow, "Introduction: Evolutionary Psychology and Conceptual Integration," in Barkow, Cosmides, and Tooby, eds., *The Adapted Mind: Evolutionary Psychology and the Generation of Culture* (New York: Oxford University Press, 1992), 3–17.
Coulter, Ann, *Godless: The Church of Liberalism* (New York: Three Rivers Press, 2007).
Coyne, Jerry, *Why Evolution Is True* (New York: Viking, 2009).
Coyne, Jerry A. and H. Allen Orr, *Speciation* (Sunderland, Mass.: Sinauer Associates, 2004).
Curd, Martin and J.A. Cover, eds., *Philosophy of Science: The Central Issues* (New York: W.W. Norton, 1998).
Dahl, Robert A., *Democracy and Its Critics* (New Haven, Conn.: Yale University Press, 1989).
———, Dahl, Robert A., *A Preface to Democratic Theory* (Chicago: University of Chicago Press, 1956).
D'Antonio, William V., Steven A. Tuch, and Josiah R. Baker, *Religion, Politics, and Polarization: How Religiopolitical Conflict Is Changing Congress and American Democracy* (Lanham, Md.: Rowman & Littlefield, 2013).
Darwin, Charles, *The Descent of Man* (New York: W.W. Norton, 1871, 2006).
———, *On the Origin of Species by Means of Natural Selection* (New York: Barnes and Noble Classics, 1859, 2004).
Davis, Michael J., "Religion, Democracy, and the Public Schools," *Journal of Law and Religion*, 25 (2009–2010), 38.
Dawkins, Richard, *The Greatest Show on Earth: The Evidence for Evolution* (New York: Free Press, 2009).
———, *The God Delusion* (Boston: Houghton Mifflin, 2006).

———, *A Devil's Chaplain: Reflections on Hope, Lies, Science, and Love* (Boston: Houghton Mifflin, 2003).

———, *Unweaving the Rainbow: Science, Delusion and the Appetite for Wonder* (Boston: Houghton Mifflin, 1998).

———, *The Blind Watchmaker: Why the Evidence of Evolution Reveals a Universe without Design* (New York: W.W. Norton, 1986, 1996).

———, *The Selfish Gene* (Oxford: Oxford University Press, 1976, 1989).

Degler, Carl, *In Search of Human Nature: The Decline and Revival of Darwinism in American Social Thought* (Oxford: Oxford University Press, 1991).

DelFattore, Joan, *The Fourth R: Conflicts Over Religion in America's Public Schools* (New Haven, Conn.: Yale University Press, 2004).

Dembski, William, "Dealing with the Backlash against Intelligent Design," in William A. Dembski, ed., *Darwin's Nemesis: Phillip Johnson and the Intelligent Design Movement* (Leicester, England: Inter-Varsity Press, 2006), 81–104.

———, "Preface" to *Darwin's Nemesis: Phillip Johnson and the Intelligent Design Movement* (Leicester, England: Inter-Varsity Press, 2006).

———, "Introduction: The Myths of Darwinism," in William A. Dembski, ed., *Uncommon Dissent: Intellectuals Who Find Darwinism Unconvincing* (Wilmington, Del.: ISI Books, 2004), xvii–xxxvii.

———, "Intelligent Design as a Theory of Information," 553–573, and "Who's Got the Magic?" 639–644, in Robert T. Pennock, ed., *Intelligent Design Creationism and Its Critics: Philosophical, Theological, and Scientific Perspectives* (Cambridge, Mass. MIT Press, 2001).

———, *Intelligent Design: The Bridge between Science and Theology* (Downers Grove, Ill.: InterVarsity Press, 1999).

Dempsey, Bernard W., "Just Price in a Functional Economy," in James A. Gherity, ed., *Economic Thought: A Historical Anthology* (New York: Random House, 1965), 4–23.

Dennett, Daniel, "Why Getting It Right Matters," in Paul Kurtz, ed., *Science and Religion: Are They Compatible?* (Amherst, N.Y.: Prometheus Books, 2003), 149–159.

Denton, Michael, *Evolution: A Theory in Crisis* (Chevy Chase, Md.: Adler and Adler, 1986).

de Tocqueville, Alexis, *Democracy in America*, Vol. I (New York: Vintaga Books, 1835, 1945).

Dowd, Maureen, "Watch Out Below!" *New York Times*, December 16, 2012, Opinion Pages.

Downs, Anthony, *An Economic Theory of Democracy* (New York: Harper and Row, 1957).

Druckman, James N., "On the Limits of Framing Effects: Who Can Frame?" *Journal of Politics*, 63, no. 4 (November 2001), 1041–1066.

Ecklund, Elaine Howard, Jerry Z. Park, and Phil Todd Veliz, "Secularization and Religious Change among Elite Scientists," *Social Forces*, 86, no. 4 (2008), 1805–1839.

Edis, Taner, "A World Designed by God: Science and Creationism in Contemporary Islam," in Paul Kurtz, ed., *Science and Religion: Are They Compatible?* (Amherst, N.Y.: Prometheus Books, 2003), 117–125.

Editorial, "Elect the Capable, Shun Ideologues," *Austin American-Statesman*, October 24 (2012), A10.

Editors of *First Things*, "Introduction" to *The End of Democracy*, Mitchell S. Muncy, ed. (Dallas: Spence Publishing Company, 1997), 6–7.

*Edwards v. Aguillard*, 482 U.S. 578 (1987).

Ellison, Christopher G. and Marc A. Musick, "Southern Intolerance: A Fundamentalist Effect?" *Social Forces*, 72, no. 2 (December 1993), 379–309.

Ellison, Christopher G. and Kathleen A. Nybroten, "Conservative Protestantism and Opposition to State-Sponsored Lotteries: Evidence from the 1997 Texas Poll," *Social Science Quarterly*, 80, no. 2 (June 1999), 356–369.
*Engel v. Vitale*, 370 U.S. 421–432 (1962).
*Epperson v. Arkansas*, 393 U.S. 97 (1987).
Evans, E. Margaret, "Cognitive and Contextual Factors in the Emergence of Diverse Belief Systems: Creation versus Evolution," *Cognitive Psychology*, 42 (2001), 217–266.
———, "Beyond Scopes: Why Creationism Is Here to Stay," in Karl S. Rosengren, Carl N. Johnson, and Paul L. Harris, eds., *Imagining the Impossible: Magical, Scientific, and Religious Thinking in Children* (Cambridge: Cambridge University Press, 2000), 305–333.
*Everson v. Board of Educ.*, 330 U.S. 1, 67 S. Ct. 504, 91 L.Ed. 711 (1947).
Faulkner, Danny, *Universe by Design: An Explanation of Cosmology and Creation* (Green Forest, Ark.: Master Books, 2004).
Ferguson, Thomas and Joel Rogers, *Right Turn: The Decline of the Democrats and the Future of American Politics* (New York: Hill and Wang, 1986).
Ferrari, Michel and Roger S. Taylor, "Teach the Demarcation: Suggestions for Science Education," in Taylor and Ferrari, eds., *Epistemology and Science Education: Understanding the Evolution vs. Intelligent Design Controversy* (New York: Routledge, 2011), 271–285.
Ferris, Timothy, *The Science of Liberty: Democracy, Reason, and the Laws of Nature* (New York: HarperCollins, 2010).
Feyerabend, Paul, *Against Method* (London: Verso, 2010).
Fodor, Jerry and Massimo Piatelli-Palmarini, *What Darwin Got Wrong* (New York: Picador, 2011).
Forrest, Barbara, "The Wedge at Work: How Intelligent Design Creationism Is Wedging Its Way into the Cultural and Academic Mainstream," in Robert T. Pennock, *Intelligent Design Creationism and Its Critics: Philosophical, Theological, and Scientific Perspectives* (Cambridge, Mass.: MIT Press, 2001), 5–53.
Forrest, Barbara and Paul R. Gross, *Creationism's Trojan Horse: The Wedge of Intelligent Design* (Oxford: Oxford University Press, 2004).
Frank, Thomas, *What's the Matter with Kansas? How Conservatives Won the Heart of America* (New York: Henry Holt and Company, 2004).
Frankfort-Nachmias, Chava and David Nachmias, *Research Methods in Social Science*, 4th ed. (New York: St. Martin's Press, 1992).
Fraser, James W., *Between Church and State: Religion and Public Education in Multicultural America* (New York: St. Martin's Press, 1999).
Freeman, Patricia K. and David J. Houston, "The Biology Battle: Public Opinion and the Origins of Life," *Politics and Religion*, 2, no. 1 (April 2009), 61–68.
Friedman, Lawrence M., *American Law in the 20th Century* (New Haven, Conn.: Yale University Press, 2002.
Futuyma, Douglas J., *Science on Trial: The Case for Evolution* (Sunderland, Mass.: Sinauer Associates, 1995).
Gay, Peter, *Modernism: The Lure of Heresy* (New York: W.W. Norton, 2008).
Ghiselin, Michael T., *The Economy of Nature and the Evolution of Sex* (Berkeley: University of California Press, 1974).
Gibbs, Jr., Raymond W., *The Cambridge Handbook of Metaphor and Thought* (Cambridge: Cambridge University Press, 2008).

———, "Metaphor and Thought: The State of the Art," in Gibbs, ed., *The Cambridge Handbook of Metaphor and Thought* (Cambridge: Cambridge University Press, 2008), 3–13.

———, *The Poetics of Mind: Figurative Thought, Language, and Understanding* (Cambridge: Cambridge University Press, 1994).

Gilbert, Scott F., "The Aerodynamics of Flying Carpets: Why Biologists Are Loathe to 'Teach the Controversy,'" in Nathaniel C. Comfort, ed., *The Panda's Black Box: Opening Up the Intelligent Design Controversy* (Baltimore: Johns Hopkins University Press, 2007), 40–62.

Gilens, Martin and Benjamin I. Page, "Testing Theories of American Politics: Elites, Interest Groups, and Average Citizens," *Perspectives on Politics*, 12, no. 3 (September 2014), 564–581.

Gilkey, Langdon, *Creationism on Trial: Evolution and God at Little Rock* (Charlottesville: University Press of Virginia, 1985).

Gish, Duane T., *Dinosaurs by Design* (Green Forest, Ark.: Master Books 1992).

———, *Evolution? The Fossils Say No!* 3rd ed. (San Diego: Creation-Life Publishers, 1979).

Gleick, James, "At the Beginning: More Things in Heaven and Earth," in Bill Bryson, ed., *Seeing Further: The Story of Science, Discovery, and the Genius of the Royal Society* (New York: HarperCollins, 2010), 27–28.

Gould, Stephen Jay, *The Structure of Evolutionary Theory* (Cambridge, Mass.: Harvard University Press, 2002).

———, *The Lying Stones of Marrakech: Penultimate Reflections in Natural History* (New York: Harmony Books, 2000).

———, *Rocks of Ages: Science and Religion in the Fullness of Life* (New York: Vintage, 1999).

———, *The Mismeasure of Man*, 2nd ed. (New York: W.W. Norton, 1996).

———, "Kropotkin Was No Crackpot," in *Bully for Brontosaurus: Reflections in Natural History* (New York: W.W. Norton, 1991), 325–339.

———, "Not Necessarily a Wing," in *Bully for Brontosaurus: Reflections in Natural History* (New York: W.W. Norton, 1991), 139–151.

———, *Hen's Teeth and Horse's Toes: Further Reflections on Natural History* (New York: W.W. Norton, 1983).

———, *The Panda's Thumb: More Reflections in Natural History* (New York: W.W. Norton, 1980).

———, *Ever since Darwin: Reflections in Natural History* (New York: W.W. Norton, 1977).

Greenawalt, Kent, *Does God Belong in the Public Schools?* (Princeton, N.J.: Princeton University Press, 2005).

Greenberg, Daniel S., *Science, Money, and Politics: Political Triumph and Ethical Erosion* (Chicago: University of Chicago Press, 2001).

Greene, William H., *Econometric Analysis* (Boston: Prentice Hall, 2012).

Gregg, Benjamin, *Human Rights as Social Construction* (Cambridge: Cambridge University Press, 2012).

Grodzins, Morton, *The American System* (Chicago: Rand McNally, 1966).

Habibi, Shahrzad, *God, School, and the Fourth R: Religion Policy in Texas Public Schools*, unpublished undergraduate honors thesis, University of Texas at Austin, (2005).

Haldane, David, "Believing's Seeing, Even in Chocolate," *Austin American-Statesman* (December 22, 2006), A27.

Hamilton, Alexander, John Jay, and James Madison, *The Federalist Papers* (New York: Modern Library, 1937), 141.
Hamilton, Marci A., *God vs. the Gavel: Religion and the Rule of Law* (Cambridge: Cambridge University Press, 2005).
Hanson, Thor, *Feathers: The Evolution of a Natural Miracle* (New York: Basic Books, 2011)
Harre, Rom, *Varieties of Realism: A Rationale for the Natural Sciences* (Oxford: Basil Blackwell, 1986).
Hart, Roderick P. and Suzanne M. Douthton, *Modern Rhetorical Criticism*, 3rd ed. (Boston: Pearson, 2005).
Hempel, Lynn M. and John P. Bartkowski "Scripture, Sin and Salvation: Theological Conservatism Reconsidered," *Social Forces*, 86, no. 4 (June 2008), 1647–1674.
Hesse, Mary B., *Models and Analogies in Science* (Notre Dame, Ind.: Notre Dame Press, 1966).
Himmelfarb, Gertrude, *Darwin and the Darwinian Revolution* (Chicago: Ivan R. Dee, 1996).
Hobbes, Thomas, *Leviathan* (New York: Collier Books, 1962).
Hofer, Barbara K., Chak Fu Lam, and Alex DeLisi, "Understanding Evolutionary Theory: The Role of Epistemological Development and Beliefs," in Roger S. Taylor and Michel Ferrari, eds., *Epistemology and Science Education: Understanding the Evolution vs. Intelligent Design Controversy* (New York: Routledge, 2011), 95–96.
Hofstadter, Richard, *The American Political Tradition and the Men Who Made It* (New York: Random House, 1948).
———, *Social Darwinism in American Thought* (Boston: Beacon Press, 1944, 1992).
Huizinga, Johan, *The Waning of the Middle Ages* (New York: Doubleday Anchor Books, 1954).
Humes, Edward, *Monkey Girl: Evolution, Education, Religion, and Battle for America's Soul* (New York: HarperCollins, 2007).
Humphreys, D. Russell, *Starlight and Time: Solving the Puzzle of Distant Starlight in a Young Universe* (Green Forest, Ark.: Master Books, 1994).
Inglehart, Ronald and Christian Welzel, "Changing Mass Priorities: The Link between Modernization and Democracy," *Perspectives on Politics*, 8, no. 2 (June 2010), 551–567.
Jablonka, Eva and Marion J. Lamb, *Evolution in Four Dimensions: Genetic, Epigenetic, Behavioral, and Symbolic Variation in the History of Life* (Cambridge, Mass.: MIT Press, 2005).
James, William, *The Varieties of Religious Experience* (New York: New American Library, 1958); first published 1902.
Jarboe, Jan, "Lady Bird Looks Back," *Texas Monthly*, (December 1994), 117.
Jensen, Richard, *The Winning of the Midwest: Social and Political Conflict, 1888–96* (Chicago: University of Chicago Press, 1971).
Johnson, Mark, "Philosophy's Debt to Metaphor," in Raymond W. Gibbs, Jr., ed., *Cambridge Handbook of Metaphor and Thought* (Cambridge: Cambridge University Press, 2008), 39–52.
Johnson, Phillip E., *Darwin on Trial*, 2nd ed. (Downers Grove, Ill.: InterVarsity Press, 1993).
Jost, John T., Arie W. Kruglanski, Jack Glaser, and Frank J. Sulloway, "Political Conservatism as Motivated Social Cognition," *Psychological Bulletin*, 129, no. 3 (2003), 339–375.
Jowett, Garth S. and Victoria O'Donnell, *Propaganda and Persuasion*, 4th ed. (Thousand Oaks, Calif.: Sage Publications, 2006).
Kahneman, Daniel, *Thinking, Fast and Slow* (New York: Farrar, Straus and Giroux, 2011).
Kamarck, Elaine C., *How Change Happens—or Doesn't: The Politics of U.S. Public Policy* (Boulder, Colo.: Lynne Rienner, 2013), 74.

Kaplan, Abraham, *The Conduct of Inquiry: Methodology for Behavioral Science* (Scranton, Pa.: Chandler Publishing Company, 1964).
Keats, John, "Lamia," in *Essential Keats* (New York: HarperCollins, 1987).
Keleman, Deborah, "Are Children 'Intuitive Theists'? Reasoning about Purpose and Design in Nature," *Psychological Science*, 15, no. 5 (2004), 295–301.
Kellstedt, Lyman A. and Corwin E. Smidt, "Doctrinal Beliefs and Political Behavior: Views of the Bible," in David C. Leege and Lyman Kellstedt, eds., *Rediscovering the Religious Factor in American Politics* (Armonk, N.Y.: M.E. Sharpe, 1993), 177–198.
Kelson, John, "Fiscal Cliff Chatter Is Putting Us on Edge," *Austin American-Statesman* (December 16, 2012), B1.
Kent, James, *Commentaries on American Law* (New York: Da Capo, 1826–30, 1971).
Kettell, Steve, "Has Political Science Ignored Religion?" *PS: Political Science and Politics*, 45, no. 1 (January 2012), 93–100.
Key, Valdimer O., Jr., *Southern Politics in State and Nation* (New York: Random House, 1949).
Kitcher, Philip, *Science in a Democratic Society* (Amherst, N.Y.: Prometheus Books, 2011).
———, "Utopian Genetics and Social Inequality," in *In Mendel's Mirror: Philosophical Reflections on Biology* (Oxford: Oxford University Press, 2003), 258–282.
———, "Born-Again Creationism," in Robert T. Pennock, ed., *Intelligent Design Creationism and Its Critics: Philosophical, Theological, and Scientific Perspectives* (Cambridge, Mass.: MIT Press, 2001), 263.
———, *The Advancement of Science: Science without Legend, Objectivity without Illusions* (New York: Oxford University Press, 1993).
———, *Abusing Science: The Case against Creationism* (Cambridge, Mass.: MIT Press, 1982).
*Kitzmiller et al. v. Dover Area School District et al.*, 400 F.Supp.2d 707 (2005).
Kleppner, Paul, *The Cross of Culture: A Social Analysis of Midwestern Politics, 1850–1900* (New York: Free Press, 1970).
Kropotkin, Peter, *Mutual Aid: A Factor of Evolution* (Boston: Extending Horizon Books, 1914, 1955).
Krugman, Paul, "The Fix Isn't In: Eric Cantor and the Death of a Movement," *New York Times*, June 12, 2014.
Krutch, Joseph Wood, *The Modern Temper* (New York: Harcourt, Brace, and World, 1929, 1956).
Kuhn, Thomas, "Metaphors in Science," in Andrew Ortony, ed., *Metaphor and Thought* (London: Cambridge University Press, 1979), 418–419.
Lakoff, George, *Don't Think of an Elephant! The Essential Guide for Progressives* (White River Junction, Vt.: Chelsea Green Publishing, 2004).
Lakoff, George and Mark Johnson, *Metaphors We Live By* (Chicago: University of Chicago Press, 1980).
Laslett, Peter, *The World We Have Lost: England before the Industrial Age* (New York: Charles Scribner's Sons, 1965).
Lasswell, Harold, *Politics: Who Gets What, When, How* (New York: Meridian Books, 1936, 1958).
Lawson, Anton E. and John Weser, "The Rejection of Nonscientific Beliefs about Life: Effects of Instruction and Reasoning Skills," *Journal of Research in Science Teaching*, 27 (1990), 589–606.
Layton, Azza Salama, *International Politics and Civil Rights Policies in the United States, 1941–1960* (Cambridge: Cambridge University Press, 2000).

Leech, Beth L., Frank R. Baumgartner, Jeffrey M. Berry, Marie Hojnacki, and David C. Kimball, "Organized Interests and Issue Definition in Policy Debates," in Allan J. Cigler and Burdett A. Loomis, eds., *Interest Group Politics*, 6th ed. (Washington, D.C.: CQ Press, 2002), 275–292.

Leege, David C. and Lyman Kellstedt, "Religious Worldviews and Political Philosophies: Capturing Theory in the Grand Manner through Empirical Data," in David C. Leege and Lyman Kellstedt, eds., *Rediscovering the Religious Factor in American Politics* (Armonk, N.Y.: M.E. Sharpe, 1993), 216–231.

*Lemon v. Kurtzman*, 403 U.S. 602, 612–14 (1973).

Lever, Annabelle, "Democracy and Judicial Review: Are They Really Compatible?" *Perspectives on Politics*, 7, no. 4 (December 2009), 805–822.

Levins, Richard and Richard Lewontin, *The Dialectical Biologist* (Cambridge, Mass.: Harvard University Press, 1985).

Lipset, Seymour Martin, *Political Man: The Social Basis of Politics* (Garden City, N.Y.: Doubleday, 1963).

*Lochner v. New York*, 198 U.S. 45, 25 S. Ct. 539, 49 L. Ed. 937 (1905).

Locke, John, *An Essay Concerning Human Understanding* (Oxford: The Clarendon Press, 1975); originally 4th edition, 1700.

———, *Second Treatise of Government* (Indianapolis: Hackett, 1690, 1980).

Lopez, Jose J., "Mapping Metaphors and Analogies," *American Journal of Bioethics*, 6, no. 6 (Nov./Dec. 2006), 61.

Loye, David, *Darwin's 2nd Revolution; Book I: Darwin and the Battle for Human Survival* (Pacific Grove, Calif.: Benjamin Franklin Press, 2010).

*Lyng v. Northwest Indian Cemetery Protective Association*, 485 U.S. 439, 447, 451–452 (1988).

MacCormac, Earl R., *Metaphor and Myth in Science and Religion* (Durham, N.C.: Duke University Press, 1976).

Maienschein, Jane, "What Is an 'Embryo' and How Do We Know?" in David L. Hull and Michael Ruse, eds., *The Cambridge Companion to the Philosophy of Biology* (Cambridge: Cambridge University Press, 2007), 324–341.

Malone, Bruce, *Censored Science: The Suppressed Evidence*, 2nd ed. (Midland, Md.: Search for the Truth Publications, 2010).

Malthus, Thomas R., *Principles of Political Economy Considered with a View to Their Practical Application* (New York: Augustus M. Kelly, 1920, 1951).

Martens, Allison M., "Reconsidering Judicial Supremacy: From the Counter-Majoritarian Difficulty to Constitutional Transformations," *Perspectives on Politics*, 5, no. 3 (September 2007), 447–459.

Martin, William, *With God on Our Side: The Rise of the Religious Right in America* (New York: Broadway Books, 1996).

Masters, Roger D., *The Nature of Politics* (New Haven, Conn.: Yale University Press, 1989).

Mayr, Ernst and William B. Provine, eds., *The Evolutionary Synthesis: Perspectives on the Unification of Biology* (Cambridge, Mass.: Harvard University Press, 1998).

Mazur, Allan, "Believers and Disbelievers in Evolution," *Politics and the Life Sciences*, 23, no. 2 (November 2005), 55–61.

McChesney, Robert W., *Rich Media, Poor Democracy: Communication Politics in Dubious Times* (Champaign: University of Illinois Press, 1999).

McClosky, Herbert, "Conservatism and Personality," *American Political Science Review*, 52 (March 1958), 27–45.

McConnell, Grant, *Private Power and American Democracy* (New York: Alfred A. Knopf, 1966).

*McCreary County v. ACLU*, 545 U.S. 844 (2005).

McCubbins, Matthew and Thomas Schwartz, "Congressional Oversight Overlooked: Police Patrols-Versus Fire Alarms," *American Journal of Political Science*, 28 (1984), 165–179.

McDaniel, Eric L. and Christopher G. Ellison, "God's Party? Race, Religion, and Partisanship over Time," *Political Research Quarterly*, 61, no. 2 (June 2008), 180–191.

McKelvey, Richard, "General Conditions for Global Intransitivities in Formal Voting Models," *Econometrica*, 47 (1979), 1085–1111.

Merton, Robert, *The Sociology of Science: Theoretical and Empirical Investigations* (Chicago: University of Chicago, 1973).

Meyer, Stephen C., *Signature in the Cell: DNA and the Evidence for Intelligent Design* (New York: HarperCollins, 2009).

Micklethwait, John and Adrian Wooldridge, *The Right Nation: Conservative Power in America* (New York: Penguin, 2004).

Miller, Gary and Norman Schofield, "The Transformation of the Republican and Democratic Party Coalitions in the U.S.," *Perspectives on Politics*, 6, no. 3 (September 2008), 433–450.

———, "Activists and Partisan Realignment in the United States," *American Political Science Review*, 97, no. 2 (May 2003), 245–260.

Miller, Kenneth R., *Only a Theory: Evolution and the Battle for America's Soul* (New York: Penguin, 2008).

Moore, John N., *Questions and Answers on Creation/Evolution* (Grand Rapids, Mich.: Baker Book House, 1977).

———, *Should Evolution Be Taught?* (San Diego: Creation-Life Publishers, 1974), quoted in Kitcher, Philip, *Abusing Science: The Case against Creationism* (Cambridge, Mass.: MIT Press, 1982), 187.

Moy, Timothy, "The Galileo Affair," in Paul Kurtz, ed., *Science and Religion: Are They Compatible?* (Amherst, N.Y.: Prometheus Books, 2003), 139–144.

*Mozert v. Hawkins County Board of Education*, 827 F.2d 1058 (6th Cir. 1987), cert. denied, 484 U.S. 1066 (1988).

Nagel, Thomas, *Mind and Cosmos: Why the Materialist Neo-Darwinian Conception of Nature Is Almost Certainly False* (Oxford: Oxford University Press, 2012).

Nelson, Thomas E., Rosalee A. Clawson, and Zoe M. Oxley, "Media Framing of a Civil Liberties Conflict and Its Effect on Tolerance," *American Political Science Review*, 91 (1997), 567–583.

Nemeroff, Carol and Paul Rozin, "The Makings of the Magical Mind," in Karl S. Rosengren, Carl N. Johnson, and Paul L Harris, *Imagining the Impossible: Magical, Scientific, and Religious Thinking in Children* (Cambridge: Cambridge University Press, 2000), 1–34.

Olson, Mancur, *The Logic of Collective Action: Public Goods and the Theory of Groups* (Cambridge, Mass.: Harvard University Press, 1965).

Orr, H. Allen, "The Population Genetics of Speciation: The Evolution of Hybrid Incompatibilities," *Genetics*, 139 (April 1995), 1805–1813.

Ortony, Andrew, ed., *Metaphor and Thought* (London: Cambridge University Press, 1979).

Paley, William, "Selections from *Natural Theology*," in Robert M. Baird and Stuart E. Rosenbaum, eds., *Intelligent Design: Science or Religion?—Critical Perspectives* (Amherst, NY: Prometheus Books, 2007), 79–86; first published 1802.
Pateman, Carol, "Participatory Democracy Revisited," *Perspectives on Politics*, 10, no. 1 (March 2012), 7–19.
Patterson, Robert W., "Fiscal Conservatism Is Not Enough: What Social Conservatives Offer the Party of Lincoln," *The Family in America*, 120 (Spring 2010), 115–133.
Pearcey, Nancy R., "Darwin Meets the Bearenstein Bears: Evolution as a Total Worldview," William A. Dembski, ed. *Uncommon Dissent: Intellectuals Who Find Evolution Unconvincing* (Wilmington, Del.: ISI Books, 2004).
Pinker, Steven, *The Blank Slate: The Modern Denial of Human Nature* (New York: Viking, 2002).
Plantinga, Alvin, *Where the Conflict Really Lies: Science, Religion, and Naturalism* (Oxford: Oxford University Press, 2011).
———, "Creation and Evolution: A Modest Proposal," in Robert T. Pennock, ed., *Intelligent Design Creationism and Its Critics: Philosophical, Theological, and Scientific Perspectives* (Cambridge, Mass.: MIT Press, 2001), 779–792.
———, "When Faith and Reason Clash," in Pennock, Robert T. Pennock, ed., *Intelligent Design Creationism and Its Critics: Philosophical, Theological, and Scientific Perspectives* (Cambridge, Mass.: MIT Press, 2001), 113–145.
Poppe, Kenneth, *Reclaiming Science from Darwinism: A Clear Understanding of Creation, Evolution, and Intelligent Design* (Eugene, Ore.: Harvest House, 2006).
Prindle, David F., *Stephen Jay Gould and the Politics of Evolution* (Amherst, N.Y.: Prometheus Books, 2009).
———, *The Paradox of Democratic Capitalism: Politics and Economics in American Thought* (Baltimore: Johns Hopkins University Press, 2006).
———, *Risky Business: The Political Economy of Hollywood* (Boulder, Colo.: Westview Press, 1993).
———, *The Politics of Glamour: Ideology and Democracy in the Screen Actors Guild* (Madison: University of Wisconsin Press, 1988).
Prindle, David and Brian Roberts, "Was Lewis Black Right? Public Opinion about Evolution in Texas." American Political Science Association Annual Meeting Paper, 2010. Available at: http://ssrn.com/abstract=1642448: accessed June 24, 2014.
Putnam, Hilary, "The Meaning of Meaning," in Keith Gunderson, ed., *Language, Mind, and Knowledge*, 7, *Minnesota Studies in the Philosophy of Science* (Minneapolis: University of Minnesota Press, 1975), 131–193.
Putnam, Robert D. and David E. Campbell, *American Grace: How Religion Divides and Unites Us* (New York: Simon and Schuster, 2010).
Raup, David M., *Extinction: Bad Genes or Bad Luck?* (New York: W.W. Norton, 1991).
Ravitch, Frank S., *Marketing Intelligent Design: Law and the Creationist Agenda* (Cambridge: Cambridge University Press, 2011).
Reed, John Shelton, *The Enduring South: Subcultural Persistence in Mass Society* (Chapel Hill: University of North Carolina Press, 1974).
Reiss, Michael J., "The Relationship between Evolutionary Biology and Religion," *Evolution*, 63, no. 7 (2009), 1940.
Richards, Mark J. and Herbert M. Kritzer, "Jurisprudential Regimes in Supreme Court Decision Making," *American Political Science Review*, 96, no. 2 (June 2002), 305–320.

Rokeach, Milton, *The Open and Closed Mind: Investigations into the Nature of Belief Systems and Personality Systems* (New York: Basic Books, 1960).

Rosenberg, Alex and Daniel W. McShea, *Philosophy of Biology: A Contemporary Introduction* (New York: Routledge, 2008).

Ruse, Michael, *The Evolution–Creation Struggle* (Cambridge, Mass.: Harvard University Press, 2005).

———, *Darwin and Design: Does Evolution Have a Purpose?* (Cambridge, Mass.: Harvard University Press, 2003).

———, "Methodological Naturalism under Attack," in Robert T. Pennock, ed., *Intelligent Design Creationism and Its Critics: Philosophical, Theological, and Scientific Perspectives* (Cambridge, Mass.: MIT Press, 2001), 363–385.

———, "Evolutionary Ethics: Healthy Prospect or Lost Infirmity?" in Matthew Mohan and Bernard Linsky, eds., *Philosophy and Biology* (Chicago: University of Chicago Press, 1988), 27–73.

Ruvolo, Maryellen and Mark Seielstad, "The Apportionment of Human Diversity 25 Years Later," in Rama S. Singh, Costas B. Krimbas, Diane B. Paul, and John Beatty, eds., *Thinking about Evolution: Historical, Philosophical, and Political Perspectives* (Cambridge: Cambridge University Press, 2001), 141–151.

Saad, Gad, "Evolution and Political Marketing," in Albert Somit and Steven Peterson, eds., *Human Nature and Public Policy: An Evolutionary Approach* (New York: Macmillan, 2003), 121.

Salisbury, Robert H., "An Exchange Theory of Interest Groups," *Midwest Journal of Political Science*, 13 (Spring 1969), 1–32.

*Santa Fe Independent School District v. Doe*, 530 U.S. 290 (2000).

Sarkar, Sahotra, *Doubting Darwin? Creationist Designs on Evolution* (Malden, Mass.: Blackwell Publishing, 2007).

Sartori, Giovanni, "Guidelines for Concept Analysis," in David Collier and John Gerring, eds., *Concepts and Methods in Social Science: The Tradition of Giovanni Sartori* (New York: Routledge, 2009), 97–150.

———, "The Tower of Babel," in Giovanni Sartori, F.W. Riggs, and H. Teune, eds., *Tower of Babel: On the Definition and Analysis of Concepts in the Social Sciences* (International Studies Association, Occasional Paper no. 6, University of Pittsburgh, 1975), 7–37.

Schofield, Norman, "Instability of Simple Dynamic Games," *Review of Economic Studies*, 45, no. 3 (1978), 575–594.

Schwartz, Thomas, *The Logic of Collective Choice* (New York: Columbia University Press, 1986).

Segerstrale, Ullica, *Defenders of the Truth: The Sociobiology Debate* (Oxford: Oxford University Press, 2000).

Sesardic, Neven, "Nature, Nurture, and Politics," *Biology and Philosophy*, 25 (2010), 433–436.

Shapin, Steven and Simon Schaffer, *Leviathan and the Air-Pump: Hobbes, Boyle, and the Experimental Life* (Princeton, N.J.: Princeton University Press, 1985).

Shattuck, Roger, *Forbidden Knowledge: From Prometheus to Pornography* (New York: St. Martin's Press, 1996).

Sherkat, Darren E., and Christopher G. Ellison, "The Cognitive Structure of a Moral Crusade: Conservative Protestantism and Opposition to Pornography," *Social Forces*, 75, no. 3 (March 1997), 957–982.

Shiffren, Steven H. and Jesse H. Choper, eds., *The First Amendment: Cases—Comments—Questions* (St. Paul, Minn.: West Publishing Company, 1991).

Shogan, Colleen J., "Anti-intellectualism in the Modern Presidency: A Republican Populism," *Perspectives on Politics*, 5, no. 2 (June 2007), 295–303.
Short, Roger. V., "Darwin, Have I Failed You?" *The Lancet*, 343 (February 26, 1994), 8896.
Shubin, Neil, *Your Inner Fish: A Journey into the 3.5 Billion-Year History of the Human Body* (New York: Random House, 2009).
Sinclair, A. and M. Pendarvis, "The Relationship between College Zoology Students' Beliefs about Evolutionary Theory and Religion," *Journal of Research and Development in Education*, 30, no. 2 (1997), 118–125.
Skinner, Burrhus F., *Walden Two* (New York: Macmillan, 1948).
Smocovitis, Vassiliki Betty, *Unifying Biology: The Evolutionary Synthesis and Evolutionary Biology* (Princeton, N.J.: Princeton University Press, 1996).
Somit, Albert and Steven Peterson, *Darwinism, Dominance, and Democracy: The Biological Basis of Authoritarianism* (Westport, Conn.: Praeger, 1997).
Spanier, Bonnie B., *Im/partial Science: Gender Ideology in Molecular Biology* (Bloomington: Indiana University Press, 1995), 128–133.
Spencer, Herbert, *Social Statics: The Conditions Essential to Human Happiness Specified and the First of Them Developed* (n.p.: Forgotten Books, 2012); first published 1850.
———, *The Man Versus the State* (Indianapolis: Liberty Fund, 1982); originally published 1884.
Sprinkle, Robert Hunt, "Bioethics without Analogy," in *Clinical Ethics and the Necessity of Stories: Essays in Honor of Richard M. Zaner* (Dordrecht, Netherlands: Kluwer Academic Publishers, 2010), 71–85.
———, *Profession of Conscience: The Making and Meaning of Life-Sciences Liberalism* (Princeton, N.J.: Princeton University Press, 1994).
Steen, Gerard, "The Paradox of Metaphor: Why We Need a Three-Dimensional Model of Metaphor," *Metaphor and Symbol*, 23 (2008), 213–241.
Steensland, Brian, Jerry Z. Park, Mark D. Regnerus, Lynn D. Robinson, W. Bradford Wilcox, and Robert D. Woodberry, "The Measure of American Religion: Toward Improving the State of the Art," *Social Forces*, 79, no. 1 (September 2000), 291–318.
Stein, Ben, *Expelled: No Intelligence Allowed*, documentary film, Premise Media Corporation/Rampant Films (2008).
Stenger, Victor, *Has Science Found God? The Latest Results in the Search for Purpose in the Universe* (Amherst, N.Y.: Prometheus Books, 2003).
Stone, Deborah, *Policy Paradox: The Art of Political Decision Making*, 3rd ed. (New York: W.W. Norton, 2012).
Strobel, Lee, *The Case for a Creator: A Journalist Investigates Scientific Evidence that Points toward God* (Grand Rapids, Mich.: Zondervan, 2004).
Tannahill, Reay, *Sex in History* (New York: Stein and Day, 1982).
*Texas Monthly v. Bullock*, 489 U.S. (1989).
"Texas State Board of Education Primary Delivers Upset," *Daily Kos* (March 2, 2010). Available at: www.dailykos.com/story/2010/03/03/842453/-Texas-State-Board-of-Education-Primary-Delivers-Upset: accessed July 1, 2013.
Tweney, Ryan D., "Toward a Cognitive Understanding of Science and Religion," in Roger S. Taylor and Michel Ferrari, *Epistemology and Science Education: Understanding the Evolution vs. Intelligent Design Controversy* (New York: Routledge, 2011), 197–212.
*Unlocking the Mystery of Life*, documentary film, Illustra Media (2002).

"Vatican's Top Astronomer: Teaching of Design Wrong," *Austin American-Statesman* (November 19, 2005), A9.

Wald, Kenneth D. and Clyde Wilcox, "Getting Religion: Has Political Science Rediscovered the Faith Factor?" *American Political Science Review*, 100, no. 4 (November 2006), 523–529.

Wallace, Alfred Russel, "Creation by Law" (1868), reprinted In Alfred Russel Wallace, *Natural Selection and Tropical Nature: Essays on Descriptive and Theoretical Biology* (New York: Macmillan, 1891), 141–166.

Waxman, D. and S. Gavrilets, "20 Questions in Adaptive Dynamics," *Journal of Evolutionary Biology*, 18, no. 5 (2005), 1139–1154.

Weber, Max, "Value-judgments in Social Science," in Richard Boyd, Philip Gasper, and J.D. Trout, eds. *The Philosophy of Science* (Cambridge, Mass.: MIT Press, 1991), 719–731.

*Webster's Seventh New Collegiate Dictionary* (Springfield, Mass.: G. & C. Merriam Company, 1971).

Weinberg, Steven, *Facing Up: Science and Its Cultural Adversaries* (Cambridge, Mass.: Harvard University Press, 2001).

———, *The First Three Minutes: A Modern View of the Creation of the Universe* (New York: Bantam Books, 1977).

Wenzel, Nikolai G., "Judicial Review and Constitutional Maintenance: John Marshall, Hans Kelsen, and the Popular Will," *PS: Political Science and Politics*, 46, no. 3 (July 2013), 591–598.

Werth, Barry, *Banquet at Delmonico's: Great Minds, the Gilded Age, and the Triumph of Evolution in America* (New York: Random House, 2009).

*West Virginia State Bd. of Educ. v. Barnette*, 319 U.S. 624, 63 S.Ct. 1178, 87 L. Ed. 1628 (1943).

Whitcomb, John C. and Henry M. Morris, *The Genesis Flood: The Biblical Record and Its Scientific Implication* (Phillipsburg, N.J.: The Presbyterian and Reformed Publishing Company, 1961).

White, G. Edward, "The Path of American Jurisprudence," in *Patterns of American Legal Thought* (Indianapolis: Bobbs-Merrill, 1978), 18–73.

Wilensky, Uri and Michael Novak, "Teaching and Learning Evolution as an Emergent Process: The BEAGLE Project," in Roger S. Taylor and Michel Ferrari, *Epistemology and Science Education: Understanding the Evolution vs. Intelligent Design Controversy* (New York: Routledge, 2011), 213–242.

Wilson, Edward O., *On Human Nature* (Cambridge, Mass.: Harvard University Press, 2004).

———, *Sociobiology: The New Synthesis*, 2nd ed. (Cambridge, Mass.: Harvard University Press, 2000).

———, *Naturalist* (New York: Warner Books, 1994).

Wilson, Glenn D., "A Dynamic Theory of Conservatism," in G.D. Wilson, ed., *The Psychology of Conservatism* (London: Academic Press, 1973), 259, quoted in Jost, John T., Arie W. Kruglanski, Jack Glaser, and Frank J. Sulloway, "Political Conservatism as Motivated Social Cognition," *Psychological Bulletin*, 129, no. 3 (2003), 339–375 (Wilson quoted on p. 347).

Witte, John, Jr. and Joel A. Nichols, *Religion and the American Constitutional Experiment*, 3rd ed. (Boulder, Colo.: Westview Press, 2011), 159.

Wolbrecht, Christina and Michael. T. Hartney, "'Ideas About Interests': Explaining the Changing Partisan Politics of Education," *Perspectives on Politics*, 12, no. 3 (September 2014), 603–630.

Workman, Samuel, Bryan D. Jones, and Ashley E. Jochim, "Information Processing and Policy Dynamics," *The Policy Studies Journal*, 37, no. 1 (2009), 75–92.

Yahya, Harun, *Fascism: The Bloody Ideology of Darwinism* (Istanbul: Arastirma Publishing, 2002).

Young, Robert M., *Darwin's Metaphor: Nature's Place in Victorian Culture* (Cambridge: Cambridge University Press, 1985).

*Zelman v. Simmon-Harris*, 536 U.S. 639 (2002).

# INDEX

abortion 154–5, 157
Adorno, Theodore 138
affirmative action 139
African Americans and tolerance 140–1; and politics in the South 146, 148
Agassiz, Louis 21
Ahmanson, Howard and Roberta Green 151
Alcock, John 26
Allah 58
American Civil Liberties Union 125, 127, 128, 131
American Political Science Association 158
analogy 34
Anderson, John 149
Anderson, Mike 104
Anthropological Society 21
Aristotle 44
Armstrong, Karen 65
atheism 68, 140, 152
*Authoritarian Personality, The* 138, 140

Bacon, Francis 10, 35
Barash, David 27
Barham, James 47
Barnes and Noble 8
Bayard, James 137, 142
Behe, Michael 75, 78–81, 106, 132
*Bible*, the 4, 14, 20; and biblical inerrancy 56, 68, 92, 94, 141; and the Big Bang Theory 16–17; and controversy between polygenists and monogenists 21; and "creation science," 127; and public opinion 90–1; and public schools 66, 67; and Reagan 149; and science 63; and Young Earth creationism 17, 71
Big Bang Theory 16–17
Black, Hugo 126
Bleier, Ruth 27
Bolnick, Dan x, 2
Bork, Robert 118
Boyle, Robert 15–16
Brennan, William 128
Breyer, Stephen 118–19
Brook, John Headley 61
Brooks, Cleanth 35
Brown, Jeff 133
Brown, Walt 17
Bryan, William Jennings 147
Buber, Martin 67
Budziszewski, Jay x
Bush, George W. 132, 150, 157

Calhoun, John C. 146
Campbell, David 101, 133
Carter, Jimmy 68, 132, 149
Cearley, Robert 128
Charney, Davida 35
Chinsamy, Anusuyu 104
*Christian Harvest Times* 69

Christianity 55; and biblical inerrancy 57; difficulties of research into 92, 140–1; and the Dover school board 133; evangelical 68; and Hamilton's proposals 137; and the *Mclean v. Arkansas* case 128; and modernism 64; Protestant 68, 140; and the Ten Commandments 145; unity of, shattered by Luther 63
Christian Right, the 150, 156, 157; *see also* Religious Right
Church of Jesus Christ of Latter Day Saints 121–2
Clinton, Bill 127
Cohen, H. Floris 12
Cold War, the 147–8
Collins, Francis 59, 61
Communist party 145
concepts 34
Condit, Celeste 38–9
Condorcet, Marquis de 25
conservatism, and conservatives, American 134; Christian variety of 137; mind of 138–42; multidimensionality of 142
Constitution, United States 63, 158; First and Fourteenth Amendments to 66, 75, 118, 119, 131, 146; and "institutional friction," 144; and Intelligent Design 152; and the jurisprudence of evolution 125–33, 134; and the politics of evolution 160; as portrayed in American government textbooks 117–18; and science 89
Copernicus 14, 63
Coulter, Ann 48
Coyne, Jerry 5, 59
creationists 5, 6, 33, 61; and culture war 142; evaluated 83; and Judge Jones' holding in *Kitzmiller* case 156–7; kinds of 71; opposed by educators 103; and Prindle's class survey 105–11; and public opinion 100; as revived by Johnson 152; and Scalia's dissent in *Edwards v. Aguillard* 128; Young Earth 17, 47, 60, 72
"culture war" 142

Dahl, Robert 159
Darwin, Charles 3–4, 8, 13, 38, 106, 144; and the argument from design 77; and biblical inerrancy 57; and conservative's view of society 46; and "creation science" 126; and the development of culture wars 62; and Ethnological Society 21; and eugenics 22–3; and modernism 64; and the origin of life 69; and Pat Robertson 133; and "substantive due process" Constitutional doctrine 119; use of metaphor by 39–42, 46
"Darwinism" 3–4, 9, 33; and *Edwards v. Aguillard* 130; and fundamentalist religion 69, 71; and George W. Bush 150; Johnson's critique of 151; as a religion 48, 131
*Darwin on Trial* 151
Dawkins, Richard 43, 47, 58, 60, 61; characterization by, of Intelligent Design 83; influence of, on Phillip Johnson 151; and political discourse of evolution 84
deism 68
Dembski, William 73–4, 106; and the argument from design 75–7; and Big Bang Theory 17; and the Christianization of science 74; on Phillip Johnson 152
democracy 90; American ambivalence about 116–17; and Constitutional jurisprudence 117, 119, 133–4; and *Epperson* decision 126; flaws in 158–59; and Intelligent Design 152; and multidimensional politics 143–5; and the politics of evolution 157–8, 160; subdiscourses of 158; and the teaching of biology 111; tension of, with scientific biology 90, 102
Democratic party 137, 142; during Franklin Roosevelt's administration 147; modern coalition of 149–50; and the politics of evolution in state legislatures 156; and Santorum Amendment 153; and Texas State Board of Education 154
Dennett, Daniel 60
Denton, Michael 151
Descartes, Renee 12
*Descent of Man, The* 22–3
Discovery Institute 106, 152
DNA 38, 74
Dobzhansky, Theodore 61
Dover, Pennsylvania 131–3, 134, 153, 157
*Dred Scott v. Sandford* 118
Dunbar, Cynthia 154

*Edwards v. Aguillard* 122–30, 134
Einstein, Albert 13

Ellison, Christopher 141
emergence 14
empiricism 48, 55, 74, 81
*Engel v. Vitale* 120, 121
Episcopal Church 59
Epperson, Susan 125
*Epperson v. Arkansas* 125–7
equilibrium and disequilibrium in party politics 143–4
"Establishment Clause" of First Amendment 67, 121, 132, 134
Ethnological Society 21
eugenics 22–4
Evans, E. Margaret 104
externalities 13

Faulkner, Danny 17–18
Federalist party 137
Feyerabend, Paul 15
First Amendment *see* Constitution
Fodor, Jerry 40
Fortas, Abe 125–6
Fourteenth Amendment *see* Constitution
framing 44–6, 150
Franklin, Daniel P. ix
"Free Exercise Clause" of First Amendment 121
Freeman, Gary x
Freud, Sigmund 64, 138
Frisch, Carl von 24
fundamentalist religion 61, 68; and the American South 141–2; and business wing of Republican party 156; and "creation science" 126; difficulties of doing research into 140–2; and the jurisprudence of evolution 123; and the "lust for certainty" 69; and moral absolutism 141

Galileo 12, 14–15, 63
Galton, Francis 23
Gay, Peter 64
*Genesis*, book of 18, 56–7, 127, 129; and biblical inerrancy 58, 68, 71; and Constitutional jurisprudence 125; and *Epperson* decision 126; Johnson's view of 151; and partisans of Intelligent Design 73; and theory of evolution 69, 90–1
*Genesis Flood, The* 126–7
genetics 19, 71
Ghiselin, Michael 42

Gilkey, Langdon 59
Gish, Duane 152
Goldwater, Barry 148
Gould, Stephen Jay 1–2, 5, 59, 61
Graglia, Lino ix, 118
Gregg, Benjamin ix, 24
Greenberg, Daniel 11
Grodzins, Morton 37

Hamilton, Alexander 116, 118, 145, 148, 152, 158, 159; and Christianity 137, 142; and Reagan 149
Haught, John 59
heliocentric theory 14
Himmelfarb, Gertrude 20
Hispanics 141, 142, 150
Hitler, Adolf 13
Hobbes, Thomas 15–16
Hodge, Charles 70
Holmes, Oliver Wendell 23, 66
Hoyle, Fred 16
Huizinga, Johan 62
Human Genome Project 24, 61
Hume, David 10–11
Humes, Edward 133
Huxley, Thomas 41–2

implication 12–14; and the compatibility of religion and science 58, 64; and "substantive due process" Constitutional doctrine 119
Inglehart, Ronald 64
"Intelligent Design" creationism 41, 60, 130–1; beliefs of adherents of 72–83; and *Kitzmiller* case 131–3, 157; origins of 151–2; and Prindle's class survey 105, 106; and state legislative politics 154; and "teach the controversy" 103, 105, 152; and Texas State Board of Education 154; and Texas state Republican platform 152–3; use of metaphor by partisans of 47

Jacobsohn, Gary x
Jefferson, Thomas 122, 123, 137, 158
Jesus 58, 62; and evangelical Christianity 68
Johnson, Lyndon 148
Johnson, Mark 35
Johnson, Phillip 106; attack by, on naturalism 81–2; involved in state

politics 153; as policy entrepreneur 150–1; use of metaphors by 47
Jones, Bryan x
Jones, John 132–3, 157–8

Kahneman, Daniel 57, 58
Kamarck, Elaine 155
Kansas State Board of Education 153
Kant, Immanuel 61
Kaplan, Abraham 10
Keats, John 64
Keleman, Deborah 57
Kelley, Patricia ix
Kellstedt, Lyman 94
Kennedy, John 149
Kent, Chancellor James 118
Kepler, Johannes 63
Key, V. O., Jr. 145
Keynes, John Maynard 23
Kitcher, Philip 24, 103
Koons, Robert x, 106
Kottler, Malcolm 104
Kropotkin, Peter 42–3
Krutch, Joseph Wood 64–5
Kuhn, Thomas 36
Ku Klux Klan 45

Lakoff, George 35, 45–6
Lasswell, Harold 25
Lawson, Anton 104
Leege, David 94
left-wing political values and philosophy 25–6; *see also* liberalism and liberals
"Lemon test" in Constitutional jurisprudence 124, 128, 130
Levins, Richard 12
Lewontin, Richard x, 12, 26
liberalism and liberals 139
Lin, Tse-min ix, x, 107
Lippmann, Walter 159
Lipset, Seymour Martin 145
Little Rock Ministerial Association 125
*Lochner v. New York* 118
Locke, John 35, 89, 157–8
Lorenz, Konrad 24
Lozanne, Arturo de x
Luther, Martin 63
*Lyng v. Northwest Indian Cemetery Protective Association* 122

Machiavelli, Niccolo 63
Madison, James 158

Maienshein, Jane 59
Malthus, Thomas 41
Marx, Karl 64, 137, 145
Matthews, Eda x
Mazur, Allan 94–5
McClosky, Herbert 139
McCubbins, Matthew 37
*Mclean v. Arkansas* 128, 130, 131
McLeroy, Don 154
McShea, Daniel W. 11
Mencken, H. L. 146, 147
Mendel, Gregor 3, 19
metaphor 34; as applied to Intelligent Design 83; applied to the jurisprudence of evolution 134; Behe's use of 79; "fiscal cliff" 37–8; operationalization of 36–7; pedagogical type of 35; and persuasion 39; in science 35; in *The Federalist Papers* 116; theory-constitutive type of 35; "wall of separation" in jurisprudence 124

Meyer, Stephen 73
Miller, Gary 155
Miller, Kenneth 10, 12, 59, 80, 132
Mivart, St. George 77–8
modernism 64–6; and American judiciary 121; and anti-evolution beliefs 69; and Dover school board 133; and *Epperson* decision 126; and party politics 145; and the South 146
Modern Synthesis 3–4, 19, 25, 58, 71; and *Edwards v. Aguillard* 129, 130; and the origin of life 74, 129; and Prindle's survey 106; and public opinion 102, 103, 104; in the public schools 90
monogenist theory 20–1
Morgan, Thomas Hunt 42
Morris, Henry 69, 126–7
Morton, Samuel George 21
Moses, Nancy x
*Mozert v. Hawkins County Board of Education* 123

National Center for Science Education 111
National Education Association 125
naturalism 81, 151
natural law 48, 55
natural selection, theory of 4–5, 18; and biblical inerrancy 56–7; and the

compatibility of religion and science 59–60; after Sputnik 70
Nazis 22, 24, 138
Nemeroff, Carol 57
Newton, Isaac 35, 56, 63, 65; and the 2nd Law of Thermodynamics 71
No Child Left Behind Act 153

O'Connor, Sandra Day 122
O'Daniel, W. Lee, "Pappy" 145
*Of Pandas and People* 131
Ohman, Jack 38
ontology 3, 48; and creationism 111; and modernism 65
Oppenheimer, Robert 11
*Origin of Species, The* 3–4, 18, 105, 106; and eugenics 22; and *Epperson* decision 126; use of metaphor in 33, 39, 41
Orr, H. Allen 5
Overton, William 127–8

Palevitz, Barry 59
Paley, William 75
Pandian, Jacob 60
Pangle, Lorraine x
Pangle, Thomas x
Pendarvis, Murray 104
Pennock, Robert 73
Percy, Charles 149, 148
Piattelli-Palmarini, Massimo 40
Planany, Eva 104
Plantinga, Alvin 60, 62; attack on naturalism 82; and political discourse of evolution 84
Plato 159
Pleistocene epoch 26
policy entrepreneurs 150
political parties, coalitional nature of 155
polygenist theory 20–1
public opinion about evolution 91–102
public schools 2; biology curriculum in 7; future of science education in 142; initial Protestant curriculum of 66; as target of creationists 83
Putnam, Hilary 36
Putnam, Robert 101, 133

Ravitch, Frank 75
Reagan, Ronald 148–49, 152; and modern Republican coalition 155
Religious Right 149; *see also* Christian Right

Religious Roundtable 149
Republican Party and fundamentalists 141–2; and abortion 154–5; and the Discovery Institute 152; modern coalition of 149–50, 155–6; and politics of evolution featuring, in the states 152–5; and Ronald Reagan 148; and Texas State Board of Education 154
*Reynolds v. United States* 121
rhetoric 34
Riess, Michael 111
Roberts, Brian ix, x, 96
Robertson, Pat 133
*Roe v. Wade* 118
Rokeach, Milton 139–40
Roman Catholic Church 14–15, 16, 59, 120; and close-mindedness 140; and theory of evolution 68
Roosevelt, Franklin 147
Roosevelt, Theodore 23
Rosenberg, Alex 11
Rousseau, Jean Jacques 157
Royal Society, The 10, 12, 16, 28
Rozin, Paul 57
Ruse, Michael 41, 59, 84

Sanger, Margaret 23
Santa Fe, Texas 68
Santorum, Rick 153
Sarkar, Sahotra 71, 83
Scalia, Anton 129–30, 134
Schaffer, Simon 15–16
Schwartz, Thomas 37
science definition of 55; and empiricism 81; prior to the Scientific Revolution 63; public ignorance of 92; as target of fundamentalist religion 70; in tension with democracy 90, 102
Schofield, Norman 155
Scientific Revolution 35, 63
Scopes Trial 70, 125, 147, 159
Secular Humanism 69, 72
secularism 3, 56, 63; and social peace 67
Segerstrale, Ullica 28
Shapin, Steven 15–16
Shaw, Daron x, 94
Shaw, George Bernard 23
Shelley, Mary 11
Sherkat, Darren 141
Shermer, Michael 59, 80
Shogan, Colleen 150
Short, Roger 103

Shubin, Neil x, 6
Sinclair, Anne 104
Skinner, B.F. 159
Smith, Adam 63, 65
Social Darwinism 18–20, 42
Society for the Study of Evolution 2
sociobiology 25–6
South, the, and Southerners (USA) 70, 141; and evangelical Christianity 146; party politics within 147–50; and the race problem 145–6; scholarship about 145–6
Southern Methodist University 151
Spanier, Bonnie 27
species 4, 5
Spencer, Herbert 18–20, 25, 42, 119
spin doctors 45
Sprat, Thomas 10
Sprinkle, Robert ix, 42, 44
Sputnik 70
Standard Model in physics 16–17
Stein, Ben 60
Stenger, Victor 60
Stone, Deborah 9, 38
Sumner, William Graham 19
Supreme Court of the United States 67, 120; attacked by Reagan 149; and *Edwards v. Aguillard* 128–9; and jurisprudence of evolution 121; and modern party politics 148; nominees to 118; and "substantive due process" doctrine 119

*Tammy Kitzmiller et al. v. Dover Area School District* 131–3, 152, 157
theism 68
Thomas More Law Center 131, 132
*Tiktaalik* 6
Tinbergen, Nikolaas 24
Tocqueville, Alexis de 89
Tweney, Ryan 57

unanticipated consequences 14
United Church of Christ 59, 120
United Methodist Church 59, 120
University of Chicago 6
University of Texas 2, 104

Veblen, Thorstein 24

Wallace, Alfred Russel 3–4, 8, 41
Warren, Earl 66
Warren, Robert Penn 35
Weber, Max 11
Weinberg, Steven 16, 60
Wells, H.G. 23
Welzel, Christian 64
Weser, John 104
Whitcomb, John 126–7
Wilson, Edward O. 25–6, 27, 28
*Wisconsin v. Yoder* 122–3